Principles of Plant Sciences Research

Principles of Plant Sciences Research

Kanti Prasad Nautiyal

CYBER TECH PUBLICATIONS
4264/3, Ansari Road, Daryaganj, New Delhi-110002 (India)
Ph.: 011-23244078. Fax: 011-23280028
E-Mail: cyberpublicationsdelhi@yahoo.com

PRINCIPLES OF PLANT SCIENCES RESEARCH

Kanti Prasad Nautiyal

First Edition 2009

Published by :

G.S. Rawat for **Cyber Tech Publications**
4264/3, Ansari Road, Daryaganj, New Delhi-110002 (India)
Ph.: 011-23244078. Fax: 011-23280028
E-Mail: cyberpublicationsdelhi@yahoo.com

Printed at : **Amit Enterprises**, Maujpur, Delhi, (9810878184)

Preface

Conscientiousness at global extent to systematically organize relevant research work approved in the realm of plant physiology, biochemistry & plant molecular biology in the last few decades all the way through creation of an international treatise series was initiated in the late last decade with publication of advances in plant physiology. In fact this has been made feasible absolutely due to laudable contributions from world scientists of eminence in specific fields. It is important to mention that within a short span of six years the treatise is made possible to attain an international status. The instant literature coming in innumerable journals of repute reflecting new technologies in plant sciences require to be assembled in a positive, productive and helpful manner by experts all over the globe. In this volume, efforts have been made to compile very needful review articles with a serious view to scientific future planning and for really making direct move towards indispensable research. Besides these, the book also inspires us to be conscious enough upon the two lines of investigation, practical and fundamental, which must always go hand in hand. There can never be a logical parting of these two aspects of our science. Correspondingly, there can never be a national separation of the pure physiologists from the practical physiologists. Our tasks are one and we must learn to march together in our performance! This volume appears to be unique in several ways especially by inclusion of a few but vital new sections, viz., biotechnology, gene expression and genetically engineered crops; transgenic plants physiology and biochemistry, etc.

Undersigned is hopeful that this work of him would be beneficial and informative for the readers. However complete care has been taken by the whole team in bringing out this book but still a few errors are likely to be remain. Your valuable suggestions are always welcome.

—*Author*

Preface

[illegible]

CONTENTS

1

Development of the Gulf of Khambhat

INTRODUCTION

On the 26th of September 1998 an agreement was signed between the Government of Gujarat, Water Resources Department and Haskoning, Royal Dutch Consulting Engineers and Architects for the preparation of a reconnaissance report on the construction of a dam across the Gulf of Khambhat. The reconnaissance mission was patronized by m/s Gujarat State Fertilizers Company Ltd., who financed the local currency component of Consultant's fees for the mission.

The terms of reference for the study call for an appraisal, not only of the technical viability of the construction of a dam but also to assess the benefits for various uses of the dam and closed-off reservoir. This report has been prepared on the basis of information made available during a mission carried out from the 26 th September to the 3 of October 1988. Besides Haskoning, an expert of the Dutch Public work Department "Rijkwaterstaat" and experts of Development Consultants Pvt. Ltd. from calcutta participated in the mission.

During the mission, which was guided by the Central designs Organisation of the water Resources Department, sufficient data could be obtained to form a general picture of the development potentials and technical viability of the dam construction. Although various development scenarios, related to the dam construction may be distinguished only two may clearly be considered as the most important, viz.:

- fresh water storage for the irrigation and water supply of the Saurashtra peninsula (fresh water scenario) ;
- tidal energy production.

These two scenarios, therefore, have been selected for further investigation, and relevant dam alignments have been assessed accordingly.

The main criteria for the selection of the alignments have been:

- availability of sufficient water depth for the implementation of tidal power works:
- the inclusion the Narmada river flow in the fresh water storage scheme.

A previous study (Wilson 1975) on the tidal power potentials in India, states that the implementation of a tidal power scheme north of Bhavnagar is not feasible because of insufficient water depth.

Dam construction therefore has been studied in three alternative alignments between the Bhavnagar-Ghogha area on the western bank and the Narmada river mouth only.

Dam construction costs in alignments north of Bhavnagar would be considerably lower. The same holds, however, for the benefits of such alignments. In the fresh water scenario, irrigation capacity would be reduced to about one third when Narmada is not included. Technical problems related to the construction of the power works in shallow water. It is recommended, however, not to entirely discard alignments north of Bhavnagar, because smaller scale developments investments may appear attractive in a step approach.

The data collected for the purpose of this study are considered of sufficient consistency and reliability for the purpose of the reconnaissance report. On various aspects assumptions had to be made which, eventually, may cause considerable changes of cost and benefit figures in a subsequent stage after collection of more information and after detailed analysis of available data.

With respect to the costs, the conditions of the foundation for dam construction are of utmost importance. The general information on the apparent composition of the bottom of the

Gulf of Khambhat supports, however, the assumption that the foundation conditions may not form a major constraint. Further investigations and studies of the dam foundation, in the prospective dam alignment are required to obtain sufficient information concerning: bearing capacity; deformations such as settlements; liquefaction (earth tremors); seepage etc. Information on the hydraulic parameters was examined with due care. The conditions are not extremely severe and do not form a major constraint and do not need hamper dam construction.

Various closure methods were examined and all are considered technically feasible, although further optimization in terms of construction time and costs is required since major gains can be made in this respect in a project of this scale, whereas also material resources of construction are to be studied regarding suitability and economics.

The least cost solution of the closure dam, considered in this report, has been estimated at 1950 crore rupees. This solution concerns the alignment Ghogha-Alia Bet-Hansol and includes a 65 km long rock fill dam provided with a 700 m spillway and a navigation lock. The impact of dam construction on environment, tide and coastal morphology needs further investigation. Anticipated impact on tide and morphology is minor.

The development scenarios that have been looked into are related to either tidal power production, fresh water storage or a combination of both.

DEVELOPMENT SCENARIOS

1. Tidal power production in combination with navigation, port development and the crossing of the Gulf of Khambhat.
2. Fresh water storage in combination with land reclamation, irrigation, navigation, Gulf crossing and fishery.
3. Combined development in which part of the basin is used for fresh water storage and part for tidal power generation in combination with the corresponding additional developments mentioned above.

For the evaluation of the tidal power scenario only, use is made of the results of the study of Wilson, with respect to the obtainable annual energy output. This leads to an energy production in the order of 20,000 GWh per year, in case 200 turbo generators of 9 m diameter are installed. The total investment costs of the scheme, including dam construction has preliminary been estimated at Rs. 10,000 crores (based on information received from the Central Electricity Authority). apart from the power production, the connection between Dahej and Ghogha will save transport costs between South India and the Saurashtra region. These savings are estimated at 0.9 billion ton km per year, which may double within some 15 years. This means that transport benefits will grow from 50 to 100 crores rupees per year approximately.

For the fresh water storage scenario an analysis was made of the availability of fresh water and the corresponding irrigation capacity. The net yields discharging in the Gulf of Khambhat are estimated at 14 * 10^9 m^3 per year on the average.

From this water some 5.3 * 10^9 m^3 could yearly be used for irrigation with 100 dependability and some 9.8 * 10^9 m^3 with 75 dependability.

The irrigable areas in the Saurashtra peninsula have been assumed to be located between mean sea level and 50-80 m above this level on the southern and western fringe of the peninsula.

Water will be available to irrigate some 700,000 ha of this area with 100% cropping intensity. This area could almost be doubled if 75% dependability is accepted. In addition to the irrigation the following benefits have been quantified from the fresh water scenario:

- some 65,000 ha adjacent to the Gulf of Khambhat can be reclaimed and irrigated ;
- the benefit of Gulf-crossing as indicated above;
- fish production may double from some 7000 tonnes per year under the present conditions to some 14,000 tonnes of fresh water fish production in the reservoir. Moreover, fish production in tanks and ponds in the irrigation area will developed.

- reduction of cost of drinking water supply of the Saurashtra region.

The total costs of the fresh water scenario including dam construction, land reclamation, irrigation scheme and compensation for the salt industry has been estimated at Rs 3750 crores. These costs could be reduced if a higher cropping intensity is assumed (Rs 3080 crores for 175% cropping intensity).

In a combined option, a tidal power scheme could be integrated in the above described fresh water storage scenario. If a 500 km^2 , 50 turbines (F 9m) scheme is constructed, the additional costs will be about Rs. 2460 crores and total investment would arrive at Rs. 6210 crores approximately.

The 500 Km^2 power scheme could generate some 4900 GWH per year, corresponding to three to four times the power requirement of the Saurashtra irrigation scheme. The introduction of the power scheme will however reduce slightly the agricultural benefits since part of the storage capacity of the fresh water reservoir is lost to power generation and also the area that can be reclaimed is reduced. The same holds for the benefits of the fisheries.

A very preliminary evaluation of the scenarios mentioned above indicates that the power scenario looks attractive for a tidal scheme. The fresh water storage option also has good potentials, which however need to be elaborated on the basis of further investigations of costs and benefits and a more in depth financial and economical analysis, specifically in order to arrive at the optimal solution in terms of the total area to be irrigated. Since a considerable part of the irrigation benefits in the Saurashtra area are lost, due to the relatively high costs of energy required for pumping, it is anticipated that the integration of a tidal power component in the fresh water option could be an attractive scheme. Preliminary calculations show internal rates of return for the latter option of about 10%.

Apart from the internal rate of return, the necessity of fresh water in Saurashtra area is also to be considered with due weightage. It is recommended to have this reconnaissance report followed by a (pre) feasibility phase during which the following aspects need attention:

- elaboration of agricultural and power benefits;
- further study of the conception dam designs inclusive; foundations; material resources/properties; construction costs; irrigation pumping schemes (inclusive prospective use of groundwater) and tidal power schemes;
- preliminary assessment of the impact of dam construction on tide, coastal morphology, environment and ecology;
- alternative alignments north of Bhavnagar and evaluation of smaller scale schemes.
- further study of available water quantity and quality inclusive inflow-, losses-, salinity-, health-, and pollution aspects.
- more detailed studies concerning sediment transport, salutation and useful life of the reservoir.
- reservoir operations (water levels and storage).
- outline assessment of consequential overland-, navigation-, and port development aspects taking gulf crossing into account.
- impact on the existing Dhavaran Thermal power station (beneficial effects of creating a sweet water reservoir).
- dovetailing of existing-, on going-, and shelved irrigation projects, into khambhat Gulf Development project.
- irrigability aspects (suitability of soils, leaching).
- a more detailed cost/benefit study.

The main objective of the (pre)-feasibility study should be the selection of the development scenario and of the scale of the scheme.

The duration of such a study is estimated at 10 months and a 100 monument expert input in the fields of: power, agronomy and irrigation, fishery and ecology, economy, dam design, material resources and excecutional aspects coastal engineering, tides and morphology.

It goes without saying that in a (per)-feasibility study the various aspect recommended, are to be assessed in close consultation with the competent authorities.

GULF OF KHAMBHAT

Basic Data

Topography and Bathymetry

For the purpose of the project the following information was made available:

(i) Topographic map, scale 1:250,000 with contour lines every 10m, covering the Saurashtra peninsula;

(ii) Topographic map, scale 1:15,000 with contour lines every 1m, covering the low lying strip of land adjoining the Gulf of khambhat.

This information was used to estimate the irrigable area and pumping requirements for the Saurashtra region and the reclaimable area on the fringe of the Khambhat reservoir. Bathymetric information could be obtained from admiralty charts. For the area north of Bhavnagar, however, this information is based on very ancient surveys carried out during the last century. Additional information could be obtained from more recent surveys that have been carried out recently in relation with the oil and gas explorations in the Gulf of Khambhat. Based on this additional information, the accuracy of which has been estimated at some 1.5 m, an indicative bathymetric map has been compiled by the consultant.

Satellite imageries (Landsat TM) of the Gulf of Khambhat give valuable information on relative sediment concentrations and thus on bathimetric and hydraulic features that cause erosion and sedimentation in the Gulf. Analysis of series of imageries may give clear indication of the morphological dynamics of the Gulf of Khambhat under the present condition.

Geology and Geotechnical Date

Due to the presence of oil and natural gas, the so called Khambhat is one of the best explored sedimentary basins of India. The basin is a long and narrow depression extending North-South and flanked on the east by the Deccan trap and on the West by Saurashtra trap.

The tectonic map shows north-south subsurface fault in the center of the Gulf and two east-west faults, one at the northern edge of the Gulf and one along the axis of the Narmada river.

The basin is essentially a tertiary basin. In the Gulf area the Tertiary and Quaternary rocks that overly the Deccan trap are covered with several tens of meters of recent alluvial deposits, generally consisting of silly fine sands.

Although several geotechnical surveys were made available, the data mostly referred to the inner area of the Gulf.

These borings show that the bottom of the Gulf mainly consists of fine sand. At some locations the sand is mixed with some thin layers of stiff clay.

At the Ghogha and Dahej ferry jetty sites the borings show however considerably more clay.

At the proposed dam site in the Gulf no borings were available.

In this study it is assumed that the bottom of the Gulf consists in the deeper parts mainly of fine sand.

Along the coast layers of clay have to be considered. Depending on their strength and their position, these layers may have to be removed.

Tides and Currents Water Levels

The following astronomical water levels have been taken from Admiralty chart 1486. These data are not in contradiction with the information gathered at several institutions during the reconnaissance mission.

The data estimated for Piram island are representative for the tidal boundary conditions in the entrance of the khambhat estuary.

The tidal variation is modeled as a sine-function with an amplitude equal to the half of the range at Piram Island.

Wind and Waves

The average wind conditions during the various seasons of the year are given below:

December-February	5-10 knts NE-winds; no precipitation
March-May	10-15 knts w to N-winds; no precipitation
June-September	30 knts SW-winds; rains
October-November	5-10 knts variable winds; some rains

The Gulf of khambhat is mainly exposed to S and SW-winds. The dam site near Ghogha-Dahej lies however sheltered for the severe SW-winds behind Gopnath point. The annual average of occurrence of wind speeds from SW-direction:

Windspeed (knts)	Occurrence
22.5-27.5	4.6%
27.5-32.5	2.0%
32.5-37.5	0.5%
37.5	0.2%

For the preliminary design of the dam a design wind speed of 60 knts (30 m/s) is chosen.

The fetch from Gopnath point to Dahej (proposed dam site) amounts to 100 km. The Bretschneider wave growth model leads to the following estimate for the significant wave height and the peak period of the wave spectrum at 25 m water depth.

Windspeed	60 knts	60 knts
Fetch	100 km	200 km
Hs	4.22 m	4.67 m
Tp	6.67 s	7.15 s

The annual average percentage of occurrence of waves from SW-direction based on measurements is given below.

Wave height	occurrence
6-8 ft	6.5
8-10 ft	3.8
>10 ft	1.8

These data are in line with the predictions made on the basis of the Bret Schneider formula.

However, also a 1/50 year significant wave height of 8m. was mentioned for a water depth of 15m. As this value exceeds the physical maximum for the wave height at a water depth of 15m. it is discarded.

For the design of the dam a significant wave height of 5.0 m and peak period of the wave spectrum of 7.5 s is adopted on the basis of these calculation. Extreme wind speed will also cause a wind set up in the entrance of the Gulf of Khambhat.

The wind set up for the design wind speed of 60 knts (30 m/s) a fetch of 100 km and a water depth of 25m. is estimated at 1.50m.

Hydrology and Sedimentology

Hydrological and climatological information has been gathered in view of the availability of fresh water to be stored in the closed-off reservoir; and in relation with the spillway capacity to be provided in a closure dam. Hydrological series of annual yields were obtained from CDO for:

Narmada river at Sardar Sarovar: 1915-1962

Mahi river at Kadana dam site: 1915-1962

Sabarmati river at Dharoi dam site: 1915-1962

On the basis of the hydrological series received, inflows from above mentioned rivers into the gulf are tabled on page 11. In computing the inflows, full utilization of planned quantities to be withdrawn from the respective river, has been taken into account. More elaborate studies based on hydrological series up to and included 1988 should be carried out in the recommended pre feasibility study.

Also hydrological information related to the Dhadhar river and the rivers draining the Saurashtra area is used. Design floods were obtained of all the schemes on rivers meeting the Gulf of Khambhat.

Silt rates were made available of the Sabarmati, Mahi, Narmada, Keri and Ghelo basins. The silt rates are from 0.040 to 0.058 Ham/km^2 /year for the three large basins. From the smaller Saurashtra basins more silt is transported, about 0.135 Ham/km\$^2$ /year on the average.

Assuming that some 50% of the net production of the various basins will be arrested in the reservoirs within these basins, then it can be calculated that a total silt load of some 46 x 10^6 m^3 will enter yearly in the Gulf of Khambhat. This corresponds to a sedimentation of some 2.3 cm per year in a Khambhat reservoir with an area of 2000 km^2. Comparison of the estimated silt load with the dead storage capacity of the reservoir (about 8*10^9 m^3) indicates that the useful life of the reservoir could be various centuries.

Water Quality

Some information was obtained of the quality (chemical and biochemical) of the waters of the Mahi, Dhadhar and Sabarmati rivers. During the mission no information could be gathered on the quantities of polluted water discharging in the reservoir. This subject requires special attention in a following stage of the project.

As to the salinity, this will be diluted after closure of the Gulf (in the case that the Gulf is used for fresh water storage). The time required for the reduction of the salt content of the water to an acceptable level for irrigation, depends on the relation between the volume of the reservoir and the yearly inflow of fresh water. A first estimate indicates that for the dam alignments considered in the report, this period may be some 3-5 years.

DEVELOPMENT SCENARIOS

Flood Protection and Land Reclamation

In the coastal area, bordering the Gulf of Khambhat, some 200,000 ha have been identified as areas within the tidal ingress.

Beyond this tidal ingress area some 700,000 ha are subject to inherent salinity.

The area under tidal ingress is unculturable, while about 75% of the so called inherent saline area may be considered cultivable, some 25% of which, however, is waste area. Conversion of the Gulf of Khambhat into a fresh water reservoir, would avoid the intrusion of saline water in the so called tidal ingress area, by which this area could become cultivable. However, the area that can be reclaimed in this way depends on the water level that is to be maintained in the reservoir, or, in case of "poldering", on the dike heights that are considered acceptable. Moreover, poldering will cut down the storage capacity of the reservoir.

For the evaluation of the reclamation benefits the assumptions have been made as shown in the following sketch.

For safety reason, a max. dike height of 2 m (exclusive freeboard) has been assumed. From the figure it appears that some 65,000 ha could be reclaimed.

The areas, beyond the tidal ingress area, that are subject to inherent salinity also require reclamation in order to improve cropping systems and the yields of crops. This type of reclamation, however, is directly related to the supply of fresh water for leaching. This water could be obtained from a fresh water Khambhat reservoir, however, these areas are likely to come under the proposed Narmada command or existing irrigating schemes. The reclamation of these areas is therefore not taken as a potential benefit of the dam project. Nevertheless, it is anticipated that the creation of a fresh water environment, will relieve considerably the salinity problem of the areas surrounding the Gulf of Khambhat.

Fresh Water Storage

Gujarat State has already executed a number of irrigation schemes from the major and other rivers. A number of schemes are also under execution or investigation. Therefore, major parts of catchment of all utilizable catchment areas are already under the plan-programme.

The total resources of Gujarat State, including those from Narmada river, are estimated at $26.5 * 10^9$ m^3 of utilizable waters of full development to irrigate about 39.4 lac. ha of land.

Besides this potential, the utilizable ground water resources of the state have been assessed as $12.6 * 10^9$ m^3, so that the ultimate irrigation potential amounts to some 52% of the cultural area of Gujarat.

To reach to the level of national average in terms of irrigation potential as well as to have some relief to drought prone areas, Gujarat needs about $25 * 10^9$ m^3 of additional waters. Interlinking of the Ganga and Narmada river has been identified as a long term solution for Gujarat's water requirements. On the other hand, after completion of the Narmada irrigation scheme, still $14 * 10^9$ on the average of fresh water will be spilled yearly in the Gulf of Khambhat. If only one half of this quantity could yearly be made available for irrigation, then already a considerable part of Gujarat's water requirements would be met. This is of special importance for the drought prone area of Saurashtra that has an isolated position with respect to the Gujarat's water resources and that even requires drinking water to be carried from distant places by railways, roads and other means.

WATER AVAILABILITY

A combined series has been worked out for the yearly inflows from the rivers Narmada, Mahi and Sabarmati, assuming full utilisation of the planned quantities to be withdrawn from the respective rivers.

From the table it can be derived that the average annual inflow of the three main rivers together will be $12.9 * 10^9$ m^3 of which $8.6 * 10^9$ or some 66 % will come form the Narmada River.

Apart from the inflow from the major rivers, also the yields have been analyzed of the following rivers.

River Dhadhar

River that drain Saurashtra

Bhadar

Utawali

Keri

Ghelo

Kalubhar

Bhogavo

Conservative estimated have been made, taking into account future withdrawals from these rivers. And the following net yields discharging into the Gulf of khambhat have been estimated at:

River Dhadar: $0.6 * 10^9$ m^3

Rivers draining Saurashtra: $0.5 * 10^9$ m^3

So the total annual average inflow into the Gulf is:

1. From 3 major rivers: $12.9 * 10^9$ m^3
2. From River Dhadhar: $0.6 * 10^9$ m^3
3. From Saurashtra rivers: $0.5 * 10^9$ m^3

Total: $14 * 10^9$ m^3

This quantity can only be used partly for irrigation purpose, since, besides the evaporation losses from the Gulf of Khambhat, part of the inflow will have to be spilled due to the limited storage capacity of the Gulf.

This storage capacity depends on:

(i) the location of the closure;

(ii) the maximum and minimum water levels that are accepted in the closed-off reservoir.

As to the location of the closure, it is obvious that if closure is considered for the benefit of irrigation, the dam will have to be constructed in such a way that the Narmada will debouch into the reservoir since this river is by far the most important supplier of fresh water.

With respect to the water levels in the reservoir, it is taken that the level fluctuation should not go beyond the natural tidal fluctuations. Allowing water levels above the high water spring would adversary affect the drainage conditions of the adjoining areas whereas subject areas also will have to be protected against such high water levels. Water levels lower than the low tidal levels may have an unacceptable impact on the ground water table in the adjoining areas.

In view of the above, the maximum and minimum water levels have preliminarily been assessed at HWS and LWS respectively.

The storage capacity corresponding to the favorable dam alignments and mentioned water level fluctuation amounts to some $16 * 10^9$ m^3.

Starting from this storage capacity and the 48 years series of inflows, a simple simulation has been carried out on a yearly basis, for various withdrawals from the reservoirs. (For the findings of these simulation reference is made to Annex 1.) The simulations show that a yearly quantity of about $7.5 * 10^9$ m^3 may be withdrawn (incl. evaporation) during 100% of the years, $9.5 * 10^9$ m^3 and $11.4 * 10^9$ m^3 may be withdrawn during 87,5% and 75% of the years respectively.

Evaporation loss in Saurashtra area is 2.3 m per year. South Gujarat, however, has a rate of 1.9 m/year. It is considered reasonable to assume a 2 m evaporation per year for the reservoir area. The rainfall in the reservoir area is 0.65 m per year. The balance between evaporation and rainfall is thus a loss of 1.35

m/year corresponding to some $2.2 * 10^9$ m^3 per year. So the net quantity of water available for irrigation amounts to:

$5.3 * 10^9$ m^3 with 100% dependability

$9.2 * 10^9$ m^3 with 75% dependability.

For a first evaluation of the irrigation benefits the 100% dependable availability of irrigation water will be considered.

Irrigable Area

In principle, the areas considered for irrigation are:

1. the peripheral zone, which can be reclaimed, mentioned in the foregoing paragraph;
2. the Saurashtra area which is not included in the command Area of Sardar Sarovar (Narmada) project.

Although no information is available on the suitability of the soils to reclaim, it is taken that the irrigation of these areas is more attractive than the irrigation of the Saurashtra area, since the irrigation costs for the latter area are relatively high because of the high pumping costs. For the Saurashtra area, water has to be lifted at least some 50 m in order to get a considerable area under command.

Assumed is that first the 65,000 ha low lying area adjoining the reservoir will be irrigated and that the remaining active storage water will be used to irrigate the Saurashtra area. From water requirement data, obtained from the Wadisang Irrigation System and assuming the cropping pattern indicated below, it is derived that the annual requirement amounts to 9853 m^3 per ha (unlined canal) corresponding to approximately 1 m at the head works.

Tidal Power Development

In the framework of present reconnaissance study two scenarios for tidal power generation are considered:

(i) Use of the Gulf for tidal power generation only;

(ii) Use of only a part of the Gulf as a tidal power basin. The major part of the Gulf can then be used to supply fresh water for irrigation.

In both cases, the tidal barrage should be located south of Bhavnagar, since the available depth north of the port is deemed insufficient to accommondate a tidal power station. Basically these are two basin systems: a single basin system or a multi (often two) basin system.

Multi basin systems are more expensive in construction and operation, but have the advantage of a fairly continuous power supply. Single basin systems supply intermittently in phase with the lunar cycle, and generally give the largest energy amount at the lowest cost. Integrating facilities for the tidal energy might be provided from existing electrical systems.

In this study a single basin system is considered. Feasibility and minimum cost of energy of more expensive alternatives may be assessed in future studies. The same holds for the choice between a one-way and a two way system. A two way system can operate during flood-as well as ebb-tide. Head difference over the turbines will be lower than with a one-way system. For the purpose of this study, a single-basin ebb-flow generation is chosen, since this system usually gives the largest quantities of energy at minimum cost. Optimization of tidal power generation schemes is a complex process. It depends on optimization techniques for many variables of both scheme design and operation. Energy estimates here are given without any optimization, and thus have but an indicative value.

Scenario I: Entire Gulf as a single basin; Ebb production; A study by E.M. Wilson * (1975) shows that annual energy output ranging between 6000 and 15000 GWh per annum is possible, depending on selection of the number of units.

On the basis of the expertise developed by the Central Electricity Authority in the field of tidal power, it has been suggested that the optimum installation of the Scenario I is of the order of 11,000 to 12,000 MW and the corresponding energy production at 20% load factor would be of the order of 20,000 GWh. Scenario II: It can be derived that if (for instance) 20% of the basin is used and 20% of the turbo generators and sluices is installed, than also 20% of the annual energy output may be expected, compared to the quantities.

Navigation-port Development

Under the present conditions two ports are in use in the Gulf of Khambhat area:

- the all weather direct berthing port of Bhavnagar and
- the fair weather lighter age port in Dahej.

Bhavnagar port is accessible for 4-5,000 DWT seagoing vessels. The water level in the docks are maintained with the help of a single lock-gate, that entering and leaving the jetties is possible during high water time only. Larger vessels anchor outside and lighterage takes place.

During the last five years the yearly quantity of cargo handled at Bhavnagar was some 500-600,000 tons per year of which some 90% concerns import. During the years 1987-88 more than 95% of the cargo was handled at anchorage.

Bhavnagar port is affected by siltation and the yearly dredging requirement are estimated at 1 million m^3 and 3 million m^3 to maintain the depth of the basin and the approach respectively.

Dahej is a ligherage port that mainly is used for the import of fertilizers. During the last years some 100,000 tons were handled yearly.

In conclusion it can be stated that lighterage is common practice in the Gulf of khambhat ports.

Future developments with respect to cargos and required handling facilities alon the Saurashtra coast have to be analyzed as part of an overall transport study for this region.

Gujarat State wishes to develop PIPAPAV as the main direct berthing do draft all weather port to cater to the needs of the ships up to 25,000 DWT. It is very likely, however, that Bhavnagar will remain an important region port that will serve a considerable part of Saurashtra and the central part of Gujarat State. Possible Developments with Dam Construction In the case of the construction of a dam "upstream" of Bhavnagar, the impact on this port could be three fold.

(i) Dam construction will impel the development of Saurashtra and by the production and cargo flows. The extent to which the port of Bhavnagar will play a role in the handling of the increased cargo, will depend very much on the accessibility of the port.

(ii) Dam construction will bring the east bank of the Gulf of Khambhat (Surat, Bharuch, Vadodara) closer to

Bhavnagar. The advantages of this development for Bhavnagar as port will depend again on the accessibility.

(iii) Bhavnagar and also Dahej may play an important role during the implementation of the dam.

In the case that dam construction will take place south of Bhavnagar, the impacts related to increased cargo flows, mentioned above, are similar, only the accessibility will be changed drastically.

First of all, provision will have to be made in the dam to allow ship transfer from the sea to the Khambhat reservoir.

Moreover, in the case that the reservoir is used essentially for irrigation purposes, it is obvious that just before the monsoon starts the reservoir level will be low (for instance corresponding to the low tide level) by which the accessibility of Bhavnagar port will be very limited, Therefore, it is considered unrealistic to provide locks for deep draft sea-going vessels. The present lighterage practice could continue if an anchorage position, preferably protected, is provided at the sea side of the dam in combination with a shiplock for shallow draft barges and river coasters.

The anchorage position, if close and connected to the dam, could be provided with cargo handling facilities so that direct loading and unloading without ligherage could take place. Such a solution however depends on whether or not sufficient depth can be maintained in front of the dam.

In the framework of the navigation study for the Narmada River that was carried out by Consultant, cargo flows were considered between the Central Indian Coal fields and the coastal area of Saurashtra and also between Bombay and Madhya Pradesh State. For the transport of this cargo a special shallow draft type of 2000 DWT river-coaster has been recommended.

Gulf Crossing

The Saurashtra peninsula is accessible for land borns traffic from the north only. A national highway connects the Saurashtra coastal area via Rajkot with Ahmedabad. The broad gauge railway runs from Ahmedabad to Kandla, with out entering the peninsula. In Saurashtra metre and narrow gauge railways are available. Land transport between Bombay and Saurashtra therefore has to make a detour that could be shortened considerably by crossing of the Gulf of khambhat. Such a crossing could be achieved by the introduction of ferry services and by the construction of a bridge or dam.

The introduction of ferry services between Dahej and Ghogha has been under consideration for more than a decade and a start was made with the implementation but has been interrupted for the time being. An important advantage of the ferry services is the relatively low investment that is required and the short implementation time. Disadvantages are, however, the limited capacity and the dependence on weather conditions. Assuming that a ferry vessel can make two turnaround trips per day between Dahej and Ghogha, then the capacity of a rather big 100 trucks (of 10 ton) vessel is limited to 200 trucks in both direction, corresponding to some 4,000 tons per day. As to capacity, the bridge or dam solution is equivalent and depends on the number of lanes that is provided. A two lane crossing is considered adequate. Technical feasibility studies have been carried out for the bridge construction.

Cargo Flows

The economic evaluation of traffic benefits of crossing the Gulf of Khambhat requires the projection of cargo flows between Saurashtra-Kachchh and the southern part of India. During this reconnaissance, no analysis of cargo flows and modal split could be obtained nor made. It was reported, however, that during a 1986 traffic survey a daily traffic flow of some fifteen hundred 10 ton trucks were counted between Bombay and Saurashtra in both directions. It was estimated that the average load factor amounts to 80-90%, so that the cargo flow using road transport between South India and Saurashtra could be of the order of 7 to 8 million tons per year.

Considering the industrial (cement, salt, etc.) and agricultural production of the Saurashtra region that is estimated roughly at some $10 * 10^6$ tons per year and the waterborne transport that is of the order of $5 * 10^6$ tons per year, it is felt that the above-mentioned cargo flow might be rather high. For a first estimate of the transport benefits of the Gulf of khambhat crosing, a road cargo flow of some $5 * 10^6$

tons per year has been assumed between saurashthra-Kachchh and Bharuch-Bombay-South India. This quantity could double in some 15 years if dam construction would imply a 5% yearly growth.

Multipurpose Development

In the foregoing section general development scenarios have beer, described more or less independently. In fact however, the only aspect of dam construction that can be considered irrespective the use of the closed-off reservoir, is the so-called Gulf crossing. For the rest, the aspects are related and depend essentially on whether fersh water storage or tidal power will be the main objective of the closure.

GULF OF KHAMBHAT CLOSURE

All multipurpose scenarios given in section 2.6 require closure at the Gulf of Khambhat from the tidal movement of the Arabian Sea.

The Location of the Dam

The location of the final dam is strongly influenced by the considerations of the closure operation. The following criteria guide the selection of the dam alignment:

1. Minimal length of the dam.

In view of the costs and the time needed for constructor, a dam of minimal length is attractive.

2. Storage capacity.

In order to maximize the water storage the dam location is pushed as far south as possible. Also in view of the availability of water the inclusion of the mouth of the Narmada is deemed necessary.

3. Use of natural points of support.

During construction the use of islands as fixed point can be attractive. They facilitate the location of material deposits and repair facilities in the neighborhood of the construction site. Piram Island which looks an obvious choice in this respect was abandoned in view of the 100 m deep gully with strong currents separating it from the mainland.

4. Perpendicular crossing of the gully.

The perpendicular crossing of gullies is advised in order to minimize construction time and cost. Also a more optimal current pattern is created during closure in a perpendicular crossing.

5. Depth of the final closure.

It is advised to start the closure of a basin with the shallower parts. This is easily done and prevents further scour.

The last part of the closure should preferably take place in a channel to minimize morphological changes in response to high current velocities. However in a relative shallow part the amount of material and the time needed are reduced. Therefore the final closure is planned in the channel west of Piram island.

6. Characteristics of the sea bed.

The alignment should be chosen in such a way that the best available foundation conditions are used. This improves the safety of the closure dam as well as the safety of the final dam. Also the scour resistance of certain types of soil or rock might be advantageously used. It is assumed on the basis of the available soil data that the seabed in the Ghogha-Dahej area consists of moderately fine to fine sand and that in this respect no special gains can be made by the choice of the alignment.

7. Alignment of the dam as a part of the coast

The final dam forms a part of the new coast line. The morphological effects of the dam on the coastal morphological processes has to be studied. At the first sight no differences are evident between alternatives.

8. Accessibility of the dam

The accessibility of the dam from existing infrastructure is of paramount importance of the final dam as well as for the dam under construction.

9. Location of shipping locks, hydropower and spilway

The location of these appurtenant structures is preferably chosen in a deep gully in a relatively sheltered position. Viewed in the light of these criteria three alternative alignments, all including the Narmada in the future fresh water reservoir, are studied.

The first alternative crosses straight for Ghogha to Dahej. The Narmada is closed by a dam over Alia Bet reaching the southern shore near Dahej. A special channel is dredged through the peninsula to discharge the Narmada in the basin. The spillway might be located in the old mouth of the Narmada.

The disadvantage of this alternative is that it implies two closure operations. The partial closure in the mouth of the Narmada and the final closure east of Priam island. The second alternative avids the dredging of a special channel by connecting the Ghogha-Dahej section with the dam on Alia Bet.

In order to save building material the third alternative starts from the coast south of Bhavnagar, proceeds in relatively shallow water to the south easterly direction and joins the second alignment in the middle of the mouth of the bay. The lay-out of the three alternatives is presented on the drawings included here.

The Cross Section of the Dam

The crest height of the final dam is calculated to prevent overtopping by sea waves.

The run up the wind waves is calculated by the formula:

Z = 8 Hs tan (alpha)

Where Hs = sign wave height

Alpha = outer slope

Z = 8 * 5 * 1/6 = 6.67 m.

The following results are reached for a 1:6 outer (down stream) slope in the upper part of the revetment.

MHWS c.d. + 8.90 m.

Wind set up 1.50 m.

Wave run up 6.67 m.

Crest level c.d. + 17.07 m.

As a preliminary crest level for a non overtopping dam a value of c.d. 18.0 m. is chosen which includes 0.93 m free board.

The original closure dam, a rock bund with slopes 1: 1.5 and a crest width of 10 m. forms part of the down stream shoulder of the completed dam.

After the closure a sand fill is placed against the rock bund at the reservoir (up stream) side. The crest width of the completed dam is 15 m. to accommodate road and rail.

The up stream side slope of the dam is 1:3 and is covered with a clay facing of 1 m. In the wave attack zone, a zone, a slope protection is provided (revetment).

The alignment of the final dam is to a large extent governed by layout of the closure operation of the basin. The various alternative closures are therefore described in more detail in section 3.3.

The crest level of the dam on Alia Bet is chosen at CD + 15 m. The wave attack at this dam section will be moderated by the extensive shoals lying at the southern side.

The reconnaissance report assumes that rock will be obtained from Chomardi on the western bank and from Netrang and Udvada on the eastern bank. Sand will be dredged from the bed of the gulf by large suction dredgers (hydraulic fill).

Spillway Capacity

One of the main threats to the safety of the Khambhat dam is an extreme rise of the reservoir level caused by a Narmada flood.

The Probable Maximum Flood (PMF) of the Narmada is 100,000 m^3 /s.

It is however not necessary to design the spillway for a discharge equal to the PMF.

If the spillway is designed for the safe release of the 1/100 year flood (67,000 m^3/s), the consequence of a larger flood is limited to the flooding of the reclaimed areas around the bay.

Although considerable damage will occur due to the flooding, the risk is acceptable in the opinion of the Consultants in view of the return period of 100 year.

Design flood criteria, will be evaluated in more detail in the recommended pre-feasibility study.

As a first estimate a gated spillway with a sill level of C.D.-3 M. and a width of 700 m is chosen. The effectiveness of this spillway in limiting the reservoir level during the design flood is checked by simulation.

Impacts of Closure

Environmental Impact

The closure of the Gulf of Khambhat implies the change of a salt water estuary into a fresh water reservoir.

The desalination of the water is a process that will take some time depending on the fresh water inflow and the content of the reservoir.

If complete mixing is assumed, the salinity will be reduced to 5% of the criminal value after the content of the lake has been replaced four time by the river inflow.

In this period the flora and fauna have to change to fresh water varieties. Experience has shown that this can be managed without great problems. For the areas surrounding the Gulf that are plagued by salt intrusion at this moment the change will stimulate a favorable development.

Impact on Tides

The closure of the Gulf of Khambhat will influence the tides on the Arabian Sea. In general experience shows that these changes are limited to an increase in the tidal range of approximately a few decimeters. This is due to thefact that the amount of energy formerly dissipated in the Gulf is only a fraction of the total energy dissipation by tidal currents in the Arabian Sea.

Locally at sharp discontinuities the current velocities might increase, however this in not expected in the Khambhat area.

The reconnaissance phase does not require thorough analysis of this effect a study of the tidal movements after the closure of the Gulf has to be made in a later stage.

Impact on Morphology

The closure of the Gulf changes the alignment of the coast and with it the impact of waves and current on the transport of sediment. At first sight the change is relatively limited so the study into the consequences for the sediment balance can be postponed to the pre-feasibility stage.

A central question in this study should be the total amount and the grainsize distribution of the sediment that is fed into the sea by the rivers. The sand fractionb of the total sediment inflow might be a major factor in the coastal equilibrium.

If, on the contrary, the sediment load of the rivers consists mainly of finer material the tidal currents will transport it to the deeper water of the Arabian Sea.

The preliminary impression, based on the apparent development of Mal Bank over the last hundred years, is that no large sand loads are brought into the Gulf by the rivers. In consequence the role of the Gulf as a source of sand is estimated to be limited and no serious erosion of the sea coast adjacent to the dam is expected after the closure. After closure all the sediment from the rivers will settle in the reservoir. Preliminary estimates indicate however that this, will be a limited deposition of a few centimeters per year.

Closure Techniques

General Overview

In the frame or the reconnaissance mission only the closure of the main channel between Ghogha and Dahej is studied because this closure the technical and economical feasibility of the project.

It is assumed that the smaller gullies in the Alia Bet are closed by means of hydraulic sandal prior to the final closure.

The calculations are centered on the closure of the approximately 30 km wide and 25 m deep entrance to the Gulf of Khambhat (bottom level C.D.-20m).

The current velocities during closure are calculated with the storage equation. The closure of the entrance can be executed in several ways:

1. horizontal closure
2. vertical closure
3. combined vertical and horizontal closure
4. sudden closure

As a feature common to all closures a bottom protection has to be laid to prevent excessive scour of the sandy sea bottom in the wake of the closure. Although the size and the weight of this protection is influenced by the closure method it is assumed in this report that an average width of ten times the water depth at both sides of the closure dam is sufficient to prevent scour induced soil mechanical instability of the dam.

Horizontal Closure

The simplest way is to build a rockfill dam "over the head" by means of dumping trucks. Thus the width of the entrance is gradually in a horizontal way.

During the closure the current velocities will gradually increase. The current velocity at spring tide is originally appr. 2 m/s (confirmed by measurements). The velocity grows via 5.5 m/s when the remaining width is 10 km to appr. 9m/s in the final stage of the closure.

The stone size that is stable under these current conditions grows accordingly. The stone size that is stable under these current conditions D, which is the product of the specific diameter of the stone and its specific weight under water.

The right hand side of the lower graph shows that for the last 5 km 3-6 t basalt blocks are required.

It may be clear that during these last 5 km the waterjet spouting through the closure gap attacks the bottom protection severely and causes serious scouring.

Vertical Closure

An attractive method of closure is the gradual heightening of the rockfill dam over the entire width of the entrance. The advantage of this vertical closure method is the limitation of the current velocity by the free overflow situation. In the case of the Gulf of Khambhat this situation arises when the crest of the dam rcached C.D.-1m. The velocity is limited to appr. 6.5 m/s at spring tide, However, in a later stage when the crest reached a level of C.D. + 4m. a maximal basalt stone weight of 6-10 t is required.

The first stages of a vertical closure are executed conveniently with stone dumping vessels. However, when the closure progresses the available water depth above the crests as well as the extreme current velocity limit the navigability. The only way to fully complete a vertical closure is by means of a cablewa or helicopters.

Combined Vertical and Horizontal Closure

A logical way to complete a vertical after navigability of dumping vessels has become questionable is a horizontal closure on top of the constructed sill. The level of the crest of the sill at which the closure method is change from vertical to horizontal should be the subject of an optimization study.

In this study the silt crest level is rather arbitrarily chosen at C.D.-3m. The rockfill dam is constructed "over the head" on top of the crest of the sill. The method combines the best of two worlds. A simples execution method and a limitation of velocities to 8 m/s.

The weight of the basalt blocks needed in the horizontal closure and in the crest of the sill is 3-6 t in a closure opening 20 to 7.5 km.

The maximal weight of the basalt blocks needed in the horizontal closure an in the crest of the sill is 6-10 t. in the final 7.5 km.

Sudden Closure

The aim of a sudden closure is to avoid extreme current velocities as much as possible thus limiting scouring and the need for extensive bottom protection works.

The method starts with a vertical closure up to a sill level that permits floating (transport) and sinking of the caisson on top of the sill. In this study a sill level of CD-10m. is assumed. Optimalisation of this level is important and may be evaluated in a later stage.

Subsequently, the remaining gap is closed horizontally by e.g. dump trucks until the current velocity just permits caisson operations. The positioning and sinking of the caissons requires as a first estimate a current window of 2 hours during which the velocity does not exceed 2m/s.

In case the current can be described by a sinusoidal function of time the current window is related to the max. velocity.

Conclusion on Closure

The conclusion of this section is that all methods are at a global level technically feasible.

All methods should however the subjected to further optimalisation in terms of construction time and costs as major gains can be made in a project of this size. Also a more thorough analysis of the various failure mechanisms of the closure dam should be made to ascertain its safe functioning during the closure operation.

In this study the analysis is limited to item to time number 2 the erosion of material due to flow as this failure mechanism is of principal importance for the technical feasibility of the closure.

The scour mechanism (no 4) which is very important in connection with the geotechnical stability is preliminary dealt with by choosing the average length of the bottom protection equal to 10 times the waterdepth. In further studies attention has to be given to this mechanism, as well as to the other mechanisms mentioned.

For the purpose of a first estimate of the costs of the dam in this reconnaissance study, the rockfill dam constructed as per combined vertical and horizontal closure method is chosen.

Cost of Closure

The cost estimate of the closure dam is made on the basis of quantities and unit prices. For a reconnaissance mission report a simple approach is advised.

Reliable rate analyses should be evaluated in the recommended pre-feasibility study. The scale of the Khambhat closure project is such that special construction methods and equipment may have to be designed to implement the project in a defined time frame. It is expcted, that the economies of scale of the project will lower the cost in comparison to an approach based on average unit prices. The total cost of the alternatives is estimated to be

Alternative 1	Rs 2720 Crores
Alternative 2	Rs 1950 Crores
Alternative 3	Rs 2110 Crores

The unit prices and the break-down of these estimates are given on the lowing two pages in which the costs are expressed in millions of Rs.

Cost Estimate of Cambay

Material	Unit price
Rock	210
R/m	3
Sand	27
R/m	3
Clay	50
R/m	3
Bottom protection	300
R/m	2
Caisson	90
Le	6
R/ piece Navigation lock	1000
Le	6
R/ piece	

All quantities are in 1E6 3 or 1E6 2

ECONOMIC ASPECTS OF DEVELOPMENT SCENARIOS

Tidal Power (Scenario A)

Costs

In the foregoing chapter an estimate has been given of the costs of dam construction. In the case of the "tidal power only" scenario, there is no need to include the Narmada mouth in the closure scheme, so that it suffices to implement only the first section (Ghogha-Dahej) of the alternative I alignment. The costs of this section are estimated at Rs. 1,850 crores.

CENTRAL WATER & POWER RESEARCH STATION

The Central Water and Power Research Station (CWPRS), Pune, as it is known today, was established in 1916 by the then Bombay Presidency as a "Special Irrigation Cell" with a limited mandate to modify irrigation practice to meet agricultural requirements and alter the agricultural methods to meet irrigation limitations. Recognising its role in the systematic study of various phases of water flow including floods, the institution was taken over by the Government of India in 1936.

With the dawn of independence and launching of planned development of the Nation's water resources, CWPRS became the principal central agency to cater to the R&D needs of projects in the fields of water and energy resources development and water-borne transport. Today, as a part of the Union Ministry of Water Resources, CWPRS is increasingly called upon to advise on projects in fields as diverse as river training and flood control, design and stable channels, irrigation and hydroelectric structures, harbours, waterways and coastal protection, structural design, integrity of structures, foundation engineering, utilization of soils, concrete and other construction materials, pumps and turbines, ship hydrodynamics, hydraulic design of bridges, earth sciences, reservoir competency, cooling water intakes, cooling pond efficiency, discharge of industrial effluents, and hydraulic instrumentation.

The current mandate of the institution encompasses undertaking specific research studies supported by necessary basic research. Comprehensive R&D support is offered to a variety of projects dealing with water resources, power and water-borne transport. Consultancy and advisory services are offered to the government within the sphere of its activities.

Disseminating expertise and research findings amongst hydraulic research fraternity, and promoting research activities at other institutions by imparting training to their research manpower, are also undertaken.

The solutions offered by the Research Station are based on the investigations from physical and mathematical models, field investigations coupled with desk studies or from a combination of these. The Research Station also collects prototype data on a variety of engineering, hydraulic and environmental parameters. The requirement of accurate and reliable instrumentation, data acquisition and control systems for physical model studies, prototype measurements are also met with by in-house developments. CWPRS with an interdisciplinary approach in all its activities thus represents unique services available to the country and the ESCAP region. The major clientele of CWPRS include:

Central Government Departments / Agencies State Government Departments / Agencies State Research Institutes

Port Trusts / State Port Organisations Public/private sector undertakings Municipal Corporations CWPRS campus, situated at downstream of Khadakwasla dam, near Pune, occupies an area of 180 Ha, where basic services include water re-circulation system for Physical Models, Library, Workshop, an Auditorium and housing facilities. A full-fledged Computer Centre provides necessary infrastructure for Mathematical Modelling work.

The present work at CWPRS covers major disciplines such as: Hydrology and Water Resources Analysis River Engineering Reservoir and Appurtenant Structures Coastal and Offshore Engineering Ship Hydrodynamics Hydraulic Machinery Earth Sciences Mathematical Modelling Foundation and Structures Instrumentation & Control Engineering

Organisational Setup

The CWPRS is a subordinate office of the Ministry of Water Resources. Apart from approval for release of funds under various heads, the Ministry of Water Resources administratively controls the functioning of the Institution. A Governing Council, a Technical Advisory Committee and a Budget and Programme Committee render advice to the Ministry regarding functioning of the Research Station.

Governing Council

The Governing Council functions as an overall policy making body under the Chairmanship of the Secretary, Ministry of Water Resources. The Council comprises members from the Finance and Administrative Wings of the Ministry of Water Resources, Planning Commission, User Organisations, State Governments and Non-government Officials. Apart from laying down broad policy guidelines for the Research Station, the Council monitors the overall progress and performance of the institution. Other functions of the Council include scrutiny and monitoring of expansion programmes, Annual and Five Year Plans, Budgetary allocations, creation and abolition of work disciplines, review of manpower requirements and delegation of additional powers to the Director.

Technical Advisory Committee

The Technical Advisory Committee, chaired by the Chairman, Central Water Commission, is primarily intended to assist the Governing Council in the matters of research and associated technical programmes. The Committee, inter alia, scrutinizes and recommends the expansion and research proposals under the Five Year Plans, suggests programmes for training of manpower and provides guidance in formulation of collaborative arrangements and Memoranda of Understanding with other institutions.

Budget and Programme Committee

The Budget and Programme Committee is Chaired by the Director, CWPRS, with Finance Officer, CWPRS as Member-Secretary. The Budget and Programme Committee assists the Governing Council to formulate budget proposals.

Staffing

The total staff strength of the research station is 1600. The Director is the "Head of Department" and Chief Administrative Officer is designated as "Head of Office". The research cadre of about 400 is supported by Technical, Auxiliary Technical, Administration, Accounts and Ancillary services.

Research and Development

The Omkareshwar Multipurpose Project is located on the Narmada River near Barwah town in Khandwa District of Madhya Pradesh. It is a run-of-river project 40 km downstream of the Indirasagar dam at Punasa. The project envisages power generation with installed capacity of 520 MW (8 x 65 MW) at river bed power house on the right bank and irrigation of approximately 2,83,300 hectares of land annually. The dam has a total length of 949 m and height of 64.0 m with a 570m long concrete spillway.

The spillway consists of 23 spans of 20 m width equipped with radial gates of size 20 m x 18.03 m. The crest of the spillway is at EL. 179.6 m. The spillway is designed to surplus the outflow flood of 88,315 m^3/s corresponding to the probable maximum flood with reservoir water level at EL. 199.62 m. Power intake structure has been proposed on the right bank of the dam. A headrace channel of length 243.2 m with average width of 208m has been provided for 8 penstocks of 7.66 m diameter.

A tailrace channel of length 212 m with average width of 242 m is proposed to discharge the water back in to the river. The full reservoir level and minimum draw down level of Omkareshwar Project are EL. 196.6m and EL. 193.54 m respectively. The hydraulic model studies were conducted at CWPRS for the spillway and power intake structures.

The hydraulic model studies for spillway were conducted on 1:50 scale 2-D sectional model. The studies indicated that the discharging capacity of the spillway would be adequate. The performance of the originally proposed slotted roller bucket as the energy dissipator was not satisfactory for the entire range of discharges. As the discharge intensities and velocities were high, the stilling basin type energy dissipator with the apron at EL. 156 m was considered suitable for this Project. The performance of stilling basin was found to be satisfactory up to 60% of design discharge. It was suggested that a 20 m long solid concrete apron may be laid on fresh rock, properly anchored and keyed at downstream of end sill to protect the stilling basin from undermining.

Hydraulic model studies for spillway and power intake were also conducted on 1:110 scale 3-D comprehensive model to study flow conditions, flow distribution and energy dissipation. In addition to original two divide walls provided to segregate the bays in the spillway, two more divide walls were suggested to contain the return eddies. The approach flow was uniform and tranquil without any vortices upstream of the spillway. In head race channel, a cross flow was observed in the channel towards the spillway which might cause deposition of silt near the intakes.

The sill of headrace channel was recommended to be raised to EL. 180 m and an upstream divide wall of 89m length with top level raised to EL. 196.60 m was suggested between powerhouse and spillway. Water level fluctuations of the order of 0.7 to 1.0 m were observed in the tail race channel when the spillway was operating along with powerhouse. In order to reduce these fluctuations and to prevent the entry of sediment into the tail race channel, it was suggested to extend the right training wall of the spillway. It was also suggested to dress down the ground near the toe of Mandhata Island near tail race channel area and increase the weir length to 280 m with its crest level raised to EL. 160.90 m. These measures were expected to lower the water levels in Kaveri branch and thereby improve the head for power generation.

A 1:60 scale geometrically similar model for power intake structure was constructed to study likely vortex formation at the intake, flow conditions in the head race channel and tail race channel for different combinations of power intake operation. The studies indicated that there was no air entraining vortex formation near the intakes for any possible combinations of intake operation and flow conditions along the head race channel were steady uniform. It was also observed that the velocity distribution at the weir crest provided at the downstream end of the tail race channel was fairly uniform.

EXTENSION OF GUIDE BUNDS OF PROPOSED ROAD AND RAIL BRIDGES ON KOSI RIVER, NLRMALI, BIHAR

National Highway Authority of India, New Delhi and East Central Railway, Patna, Bihar, have proposed construction of a road bridge and a rail bridge across river Kosi near Nirmali in North Bihar. Earlier, the studies were carried out at CWPRS for finalizing the locations and alignments of the bridges, design of guide bunds and effect of constricted waterway on the existing flood embankments upstream. The maximum afflux was observed to be 1.35 m and its effect was felt over a distance of 8 km upstream. Subsequently, as decided in the Technical Committee meeting, the hydraulic model studies were conducted at CWPRS for extension of upstream guide bunds to the existing flood protcction embankments to neutralize the effect of afflux during high floods.

The studies have been carried out on the physical 3-D model of Kosi river, constructed to a horizontal scale of 1/500 and a vertical scale of 1/70. The design discharge of 22,375 m^3/s was considered for the model studies. The suitable alignments and lengths of the right and left afflux bunds were recommended. The length of right side afflux bund was 7950 m and maximum velocity of flow, as observed along this bund was of the order or 0.50 m/s. The length of left side afflux bund was 8325 m and maximum velocity of flow, as observed along this bund was of the order of 0.60 m/s. The top levels of afflux bunds at various locations were worked out by adding free board of 1.5m to the water levels observed at these locations. It was recommended to protect the side slope of the afflux bunds on riverside by providing stone paving and other side of the bund with vegetal cover or turfing.

The Lower Goi Project envisages the construction of a 2211 m long earthen dam with a 175 m long concrete spillway across Goi river, a tributary of Narmada river, near Silavad village in Barwani district. The Upper Beda Project envisages the construction of a 2239 m long earthen dam with a 175 m long masonary spillway across Beda river, a tributary of Narmada river, near Nemit village in Bhikangaon Taluka of Khargone district. These two dam sites are located very close to the Barwani-Sukta fault in seismically active Son-Narmada-Tapti (SONATA) tectonic province. The Kutni Project envisages construction of a 30.36 m high 2625 m long composite dam across Kutni river, a tributary of the Ken river, near Khajua village in Chatarpur district. This dam site is located in the Bundelkhand gneissic terrain of Archaean-Proterozoic age, which is bound by the tectonic features associated with very low level of seismicity. The studies were conducted at CWPRS for estimation of the site-specific design seismic parameters for earthquake resistant design of these three projects by a comprehensive seismotectonic evaluation of the regions around the projects. The design parameters were obtained in the form of acceleration time history and the response spectra for several different damping values.

The past seismicity data in the respective project area for a period of about 200 years were correlated with the tectonic features and were also analyzed to estimate the magnitude of maximum credible earthquake (MCE). The MCE was found to be 6.5 for both Lower Goi and Upper Beda Projects, which were assumed to occur at a distance of 2 km from the site for Lower Goi Project and 7 km for Upper Beda Project with a focal depth of 25 km. The MCE for Kutni Project was found to be 6.0 to occur at a distance of 30 km with a focal depth of 15 km. Further, both the regional and the local geological conditions for all the project sites were taken to be of hard rock type. These specifications were used to obtain the 5% damped response spectra of horizontal and vertical components with a confidence level of 0.84, which provided the deterministic estimate of the MCE level of ground motion and were used to generate compatible accelerograms separately for the horizontal and the vertical components of motion, by incorporating physically realistic phase difference between the two components. The smoothed acceleration response spectra were then computed from these accelerograms for damping ratios of 1, 3, 5, 7, 10 and 15 % of critical. The zero period acceleration values which are equivalent to the peak ground acceleration (PGA) for horizontal and vertical components were estimated for these Projects. The MCE level of accelerograms and response spectra were recommended to be used to test the safety of the dam under extreme but rare load condition.

STADIUM COMPLEX ON YAMUNA RIVER BANK ALONG DELHI-NOIDA FLYWAY & ROAD BRIDGE ACROSS DOWNSTREAM OF WAZIRABAD BARRAGE, NEW DELHI

M/s RITES have proposed to construct a road bridge 800 m downstream of Wazirabad barrage in New Delhi. Also, the Delhi Development Authority (DDA), Delhi, has proposed a complex for cricket-football stadium of an area of about 85 hectares on Yamuna river bank surrounded by right approach embankment of Delhi-Noida Toll Bridge and proposed Kalindi bypass (NH 24), New Delhi. The studies were undertaken at CWPRS to examine the feasibility of these proposals from hydraulic point of view. Studies were undertaken on the existing mobile bed model of river Yamuna at Delhi (scales: 1/300 H & 1/60 V) covering the river reach of 29 km upstream and 21 km downstream of ISBT bridge. Studies were carried out under existing conditions and with proposed developments in position with three discharges namely 7,022 m^3/s, 9,910 m^3/s and 12,750 m^3/s corresponding to a maximum discharge observed in 1988 at Wazirabad barrage, the design discharge and check flood suggested by the Central Water Commission.

Studies indicated that the proposed waterway of 570 m between the abutments would be adequate and the afflux due to the proposed bridge at 800 m downstream of Wazirabad barrage would be negligible. However, in order to improve the flow conditions and to distribute the flow more uniformly along the bridge axis, 350 m long guide bund making 110° angle with bridge axis on the left upstream side and 190 m long guide bund on the left downstream side were recommended. On the right side, no guide bund was recommended. Water level, maximum velocity and maximum discharge intensity along the proposed bridge axis were also estimated for different discharges. The proposed bridge would not cause any undesirable flow conditions at Wazirabad barrage. Thus, the location of the bridge axis 800 m downstream of the Wazirabad barrage (at existing Pontoon bridge axis) was hydraulically satisfactory.

The model studies were undertaken for the proposed cricket-football stadium under existing conditions and with the bund (closing the gap between downstream right guide bund of Noida Toll Bridge and proposed Kalindi bypass road) position with different discharges. Existing river bed levels in the vicinity of the proposed stadium vary from RL 202.00 to 203.00 m. The depth of the water varies from 1.75 m to 2.75 m in the vicinity of the proposed site for the proposed stadium with the river discharge of 7,022 m^3/s. As per the model studies, the water level observed in the vicinity of the proposed stadium was RL 205.25 m with bund and RL 205.15 m without bund, corresponding to river discharge of 12,750 m^3/s. It was recommended that the proposed location of the stadium would be hydraulically suitable only after raising the entire area (approximately 85 ha) by 1.0 m above HFL (RL 205.00 m) along with proper drainage arrangement.

A road bridge has been constructed on river Gandak at Dumaria. The waterway of Dumaria Bridge is 843m having 13 spans of 61 m each and end spans of 25 m. The river Gandak has a channel width of about 250 m during lean period. Village Sallehpur is located close to the right guide bund of Dumaria Bridge and village Hasanpur, further upstream on right bank at a distance of about 6 km from the bridge. The concentration of flow of river Gandak at Dumaria Road Bridge has been observed to be shifting to the right bank in a width of about 250 m covering 2-3 spans of Dumaria Bridge. During the floods of the year 2002, as concentration of flow was towards the right bank, high pressure of flow was observed on the embankments at Hasanpur and Sallehpur and damage to the embankments and bank erosion were observed at these places. On the other hand, the large shoal of height 2.5 m to 3.0 m on the remaining left side spans of Dumaria Bridge was more or less inactive even during high floods.

CWPRS was requested to suggest anti-erosion measures for the eroding bank and to evolve long-term solution to shift the active channel away from the affected right bank. Before undertaking detailed model studies, the site inspection was carried out by CWPRS officials.

RIVER TRAINING WORKS FOR RIVER GANDAK AT DUMARIA GHAT, BIHAR

The observations made during the inspection and analysis of the available data indicated that the river training measures could be evolved in three stages, namely; "immediate" to provide protection to the eroding bank, "short-term" to initiate shifting of channels away from the right bank to the desired position, and "long-term" solutions to stabilise the channels in the upstream reach so that the problem may not re-occur. With the limited available data, the "immediate" and some of the preliminary "short-term" training measures were suggested.

Anti-erosion measures were suggested in the form of semi-permeable spurs made up of bamboo porcupines at the affected bank, as "Immediate" measure.

Excavation of pilot channel within the dead channel flowing along the right bank in the reach 7 km to 3 km upstream of Dumaria bridge and construction of semi-permeable screens across the existing active channel were suggested. It was felt that the re-activation of old course would help to relieve pressure on the right bank upstream of the bridge. The re-activation could also help to reduce the erosion between 8 km and 11 km upstream due to reduction in the discharge of the attacking channel.

The detailed morphological investigations and hydraulic model studies would be necessary to evolve the further "short-term" and "long-term" measures.

The Sardar Sarovar Project is located at 5.63 km upstream of Navagam village in Bharuch District of Gujarat. The project comprises a concrete gravity dam 1210 m long and 155 m high across the river Narmada. The project provides for a riverbed underground powerhouse of 1200 MW with 6 units of 200 MW each and also a canal head powerhouse with 5 units of 50 MW each. Two irrigation bye-pass tunnels (IBPT), each of 5.50 m dia. have been provided on the right bank of the river to pass water into Pond of canal head power house which would further bifurcate into two tunnels of 2.6 m x 3.9 m size at its exit. The total length of the tunnels, including bifurcation, would be 241.23 m. These tunnels were provided for supply of irrigation flows to Gujarat and Rajasthan to meet their irrigation requirements, especially during peak withdrawl for irrigation and to bye pass the flows to facilitate maintenance of turbines of canal head powerhouse. Each irrigation tunnel was designed to discharge 316 m^3/s for reservoir water level at EL. 138.68 m (FRL) and downstream maximum pond level of EL. 95.10 m. The studies were carried out at CWPRS for assessing the discharging capacity of IBPT and hydraulic downpull and uplift forces on the emergency gate using the 1: 25 scale geometrically similar model.

The energy dissipation arrangement downstream of each IBPT tunnel would be provided with independent stilling basins, each of 72.88 m length and 18.8 m width with an apron. Water from stilling basins would be discharged into downstream pond whose minimum and maximum water levels correspond to EL. 92.96 m and 95.10 m respectively. The water from pond No.1 would then lead to the Narmada canal through a series of interconnected ponds. Fixed wheel emergency gate of size 4.32 m x 5.50 m is provided in each tunnel at distance of about 58.7 m from its entrance. Two vertical slide type service gates of size 2.6 m x 3.9 m are provided at the exit of each tunnel bifurcation.

The discharging capacity of the irrigation bye-pass tunnel was measured for various reservoir water levels between EL. 88.39 m and EL.124 m and the rating curve was suggested for IBPT canal before bifurcation. The model studies on emergency gate, for water level corresponding to EL.110.64 m (MDDL), EL.116 m, EL.124 m and EL.138.68 (FRL) for assessing the hydrodynamic downpull or uplift forces indicated that the emergency gate would experience maximum net hydrodynamic down pull force of the order of 28 tonnes, 30 tonnes, 32 tonnes and 40 tonnes respectively at about 75 % gate opening and maximum net uplift force of the order of 22 tonnes, 25 tonnes, 28 tonnes and 39 tonnes respectively at about 3% gate opening for above reservoir water levels. The net uplift and downpull forces would be considered to decide hoist capacity and the self weight of the gate for self closure.

Sardar Sarovar is prestigious multipurpose project across river Narmada to cater the requirements of about 18,75,000 ha in Gujarat and Rajasthan states. Narmada Main Canal (NMC) with design discharge of 1133 m^3/s at the head regulator, traverses through Gujarat state and crosses many rivers including Mahi, Sabarmati, Meghana, Watrak, Banas, Khari-II and Saraswati.

The rivers Banas, Khari-II and Saraswati flow through the state of Gujarat and dissipate into Little Rann of Kutchh. The general topography of the rivers is very flat and water enters flood plains even under normal floods. The shifting of channels is noticed in Banas River during the past decades.

No gauging site is located on any of the rivers within the study area. M/s Sardar Sarovar Narmada Nirman Ltd. (SSNNL) faced difficulty in computing high flood levels in these three rivers due to flat terrain, large water spread and braiding pattern.

The waterways of the rivers Banas, Khari II and Saraswati were finalized during the 38th meeting of Board of Consultants, wherein representatives from CWPRS, IRI Roorkee and SSNNL were present. It was also decided in the meeting that there would be only one structure across Totana arm of Banas river. The studies for estimation of design parameters of syphons at rivers Banas, Khari-II and Saraswati crossing with NMC were carried out at CWPRS for different combinations of guide bund, afflux bunds and marginal bund.

The studies were conducted for estimation of water level, peak flood and velocity at NMC crossing during the passage of Probable Maximum Flood (PMF) and Standard Project Flood (SPF). The present studies were carried out using 1-D mathematical model (NETWORK) and the Topographic data of the river etc.

The parameters like spacing and alignment of bunds for river Banas were supplied by SSNNL. The results for river Banas were reviewed and details like spacing of afflux bund for rivers Khari-II and Saraswati were derived.

ESTIMATION OF DESIGN PARAMETERS OF NMC SYPHONS ON RIVER BANAS, KHARI-II AND SARASWATI, GUJARAT

The studies have been carried out in three stages. In the first phase, sensitivity analysis for spacing of marginal bunds was carried out. In the second phase, flows with straight guide and afflux bunds were simulated for Banas-Khari-II system using different combinations of SPF and PMF. The studies were carried out under existing conditions and lowering the bed to reproduce the desired scour. In the third phase, unsteady flow simulations were carried out to estimate design parameters viz. water level, velocity at NMC crossing for PMF, SPF and 100 year floods for elliptical guide bunds and afflux bunds with flaring. The values of parameters to be adopted in the river training and bank protection works as well as deciding top of NMC embankment at crossing were recommended.

To exploit the full potential for irrigation and power generation of the Narmada River Basin (NRB), it is proposed to construct 30 major, 135 medium and about 3000 minor dams along the main river as well as on its tributaries. The Sardar Sarovar Project (SSP) in Gujarat would be the terminal dam of this chain. In addition to the generation of 1450 MW of power, the SSP envisages a large irrigation canal system consisting of the Narmada Main Canal, branch canals, distributaries, minors and subminors. The length of Narmada Main Canal would be about 468 km with discharging capacity of about 1134 m^3/s at its head. There will be in all 42 branch canals off-taking the Narmada Main Canal, out of which Kutch Branch Canal (KBC) is one of the major branches.

ESTIMATION OF SITE-SPECIFIC DESIGN PARAMETERS AND LIQUEFACTION POTENTIAL OF KUTCH BRANCH CANAL, GUJARAT

Susceptible to liquefaction failure due to earthquake. The studies were carried out at CWPRS to assess site specific design seismic parameters for earthquake resistant design of earthen embankments of KBC and liquefaction potential assessment of the foundation strata of this canal.

The past earthquake data, geological and tectonic features in the region were analyzed. For this purpose, two critical sites have been identified where the canal has the maximum filling (designated as Site-1) and where it has the maximum cutting (designated as Site-2). In view of large separation between these two sites, the design ground motion was estimated separately. To estimate the design ground motion, both the deterministic and the probabilistic approaches were adopted in the study. The deterministic approach is based on the specifications of a Maximum Credible Earthquake (MCE) whereas the probabilistic approach takes into account the effects of total expected seisimicity in the area around the site of interest. To obtain the design response spectra, a suitable The Kutch Branch Canal (KBC) off taking from Ch. 385.814 km of Main Canal is a major branch canal having total length of 352 km with 170 m^3/s discharge capacity. The KBC would lie in the Kutch area of Gujarat, which is characterized by high level of seismicity. The entire area is traversed by several E-W trending active faults, such as Kathiawar, Kutch Mainland, Island Belt and Allah Bund faults. The region is known to have experienced many large and moderate earthquakes in the past. The proposed canal would be passing through the Bhuj area, which has been the seat of M7.7 earthquake of 26 January 2001.

The seismic safety of the KBC is therefore very important. The major length of this canal is in the seismic zone V and the foundation strata mainly consist of silty fine sand / sandy silt which are frequency-dependent attenuation relationship was identified from the published literature, using the available ground-motion data from the main Bhuj earthquake and nine of its major aftershocks. The deterministic design response spectra were found to be higher than the corresponding probabilistic one, and hence the same were recommended in view of the safety of the canal embankments. Design accelerograms compatible with these spectra were generated using random phases. The acceleration response spectra for different damping ratios, as obtained from the design accelerograms, were also found for carrying out simplified response analysis of the canal embankments.

Liquefaction analysis for the foundation strata of the KBC portion from 20.0 km to 112.5 km was carried out using the Simplified method as proposed by Seed & Idriss. In this method, Cyclic Stress Ratio (CSR) induced in the foundation soil due to the peak ground acceleration (PGA) is compared depth-wise with the Cyclic Resistance Ratio (CRR) which is computed from SPT values. The earthquake magnitude of 7.8 with PGA of 0.344 g was considered in the present analysis. The analysis indicated that some regions along alignment of canal are susceptible to liquefaction for an earthquake of magnitude 7.8 with maximum depth of liquefaction of 12.0 m below ground level in Rann of Kachch (at CH.86.0 km).

The North Eastern Electric Power Company (NEEPCO) has proposed to install a 280 MW gas based power project near Sonamura town about 65 km south-west of Agartala, Tripura. The generation capacity would be increased to 500 MW in the 2nd stage. The make-up water requirement for ultimate generation would be 0.31 m^3/s which was proposed to be drawn from Gumti river located at about 5 km from plant site. On the basis of the site inspection and discussions, the desk studies for various aspects as below were carried out at CWPRS:

Deciding location of intake and its hydraulic design including necessary protection works Examining necessity of desilting basin, its location and design Detailed design of pump-sump and pumping system including pipeline.

INTAKE WELL AND DESILTING BASIN FOR TRIPURA GAS BASED POWER PROJECT AT MONARCHAK, TRIPURA

August 1983, the corresponding water level was 14.65m. Based on gauge-discharge curve developed from observed data, the water level corresponding to 100 year return period flood (1320 m^3/s) and design discharge of 2468 m^3/s adopted for Sonamura bridge have been estimated as 15.20m and 15.60 m respectively. Analysis of sediment data indicated high sediment concentration between 1500 ppm to 4700 ppm during flood period extending over 3-11 days.

The coarse and medium sediment would be about 30% which was required to be minimized from pump operation and maintenance considerations. Considering this, an open intake with a provision of 10m long low level weir with crest at 0.50 m below minimum water level and 12 x 10m silt trap with sill level of 4.0 m below minimum water level was recommended.

Bank protection works for a distance of 30 m upstream and 10m downstream comprising wooden piles / steel piles for toe protection and stones in crate laid over synthetic filter for slope protection was suggested. Considering various aspects such as high bank levels, ease of construction and removal of deposited sediment and operation and maintenance of pumps and pipeline etc., the desilting basin was proposed to be located at the plant site.

Two units of desilting basin with one stand-by were proposed. One unit would be in operation while the other could be cleaned/maintained. As manual removal of settled sediment was proposed, no discharge was required to be pumped for flushing of the settled sediment. Twin type desilting basin, having each unit 30.0 m long and 3.0m wide, was designed for 90% settlement of sediment having diameter 0.10 mm and above with inlet discharge of about 0.32 m^3 /sec. This size was adequate for more than 90% settlement of suspended sediment coarser than 0.10 mm. After due consideration of various aspects and frequency of clearing the basin, the Project Authorities were of the opinion that the frequency of The intake location was finalized at about 700 m upstream of bridge at Sonamura. The deep channel of the river was close to the left bank and had remained there for last 20-30 years and ,moreover, at this location, the river bank was relatively steep and stable.

The minimum flow of the order of 2.80 m^3/s in river Gumti was observed with corresponding water level of 8.35m at the intake site with river bed level as about 7.20 m. For maximum observed flow of 993 m^3/s in cleaning the basin should be reduced by providing large trap capacity basins.

With the maximum monthly average inlet concentration of 1.918 gms / litre and 60% (assumed) overall settling efficiency of the desilting basin, about 681 cum of sediment would settle in the desilting basin during a month's period. The alternative proposal of the desilting basin consisted of an inlet tank 3 m wide x 3 m long to accommodate two inlet pipes, grid wall, 2.00 m wide channel followed by 1 m channel leading to each unit of desilting basin, control gate, 20 m long inlet transition, 20 m long x 9 m wide basin, and a weir directly discharging into the Raw Water Reservoir. The depth of the trench was recommended to be 2.30 m at the beginning varying to 2.70 m in the length of 32 m, thus, providing about 660 cum capacity to accommodate settled sediment which was adequate. The overall settling efficiency of this proposal was estimated analytically, using Camp's criteria, which worked out to be 79.60 %.

The Koyna dam, a multipurpose project for irrigation and hydroelectric power generation, is 103 m high and 807.7 m long across river Koyna, in the Satara District of Maharashtra State. Strengthening of the overflow sections of the dam comprising block nos. 18 to 24 to augment present storage capacity of dam by 1.5 meters and also as a precautionary measure against earthquake, has been taken up by the project authorities. The project authorities planned to complete the strengthening work in two working seasons. The strengthening work comprised placing of concrete in two stages-strengthening concrete and closure concrete. The strengthening concrete was being placed leaving a gap of 1.22 m from face of old main dam. Concrete would be placed in lift height of 1.5 m and lift interval of 72 hrs.

The closure concrete in the gap of 1.22 m was placed 30 days after the strengthening concrete from the consideration of shrinkage and thermal effect. In such mass concrete structures, estimation of suitable placement temperature on the basis of thermal, strength and elastic properties, to limit the maximum temperature rise to avoid cracking, is important. The estimated placement temperature of concrete mix is generally below the average ambient temperature. Hence the concrete mix is required to be precooled. Further in the case of such strengthening works where closure concrete is placed subsequently to join strengthening concrete with old concrete, it is also ensured that the tensile stresses due to shrinkage of strengthening concrete are minimized as far as possible within scheduled time (30 days in this case). This minimizes probability of separation at the interfaces of adjoining concrete. The shrinkage (tensile) stresses can be minimized by postcooling the concrete mass at a faster but restricted rate (0.15° C per hour). This is achieved by embedding cooling pipe meshes between each lift and circulating cold water through these pipes at predetermined temperature and at predetermined rate of flow. The studies were undertaken at CWPRS to determine the thermal properties, strength and elastic properties of strengthening concrete and closure concrete, to suggest the suitable placement temperature and cooling arrangements.

The cement level of strengthening and closure concrete were 177 kg/cum and 200 kg/cum respectively. The laboratory studies indicated 90 days compressive strength of 458 and 495 kg/cm^2, flexural strength of 49 and 50.3 kg/cm^2, specific creep of 2.32×10^{-6} and 2.80×10^{-6} $mm/mm/kg/cm^2$ and ultimate strain capacity of 226×10^{-6} and 250×10^{-6} respectively. The thermal properties of concrete mixes for strengthening and closure concrete indicated adiabatic temperature rise of 26.6° C and 29.9° C, diffusivity of 0.0027 m^2/hr. coefficient of thermal expansion 7.59×10^{-6} and 7.61×10^{-6} mm/mm/0 C respectively. On the basis of the computations, the suitable placement temperatures were estimated to be 18° C to 20° C for the strengthening concrete and 16° C to 18° C for closure concrete. In order to cool and to minimize the stresses due to shrinkage, postcooling of strengthening and closure concrete was suggested. For this purpose, GI pipe mesh of 25 mm diameter at the bottom of first lift and interface of each lift was placed. It was recommended to circulate cold water at temperature of about 22° C for a period of 25 to 30 days through the pipe mesh. The pre-cooling and postcooling respectively helped to restrict the temperature rise and to lower the temperature of concrete mass within desired period.

In order to achieve optimum utilization of water from Koyna reservoir, it was proposed to lower the intake level of state-IV by 12 meters to obtain an additional supply of 15 tmc of water. The proposed scheme included construction of an additional Surge Gallery of size 75 m(L) x 20 m(W) x 35 m(H), as well as 4.5 km long Horse Shoe shaped Head Race Tunnel, Intake Structure Complex and Lake Tapping.

CWPRS undertook field studies to determine in-situ rock mass properties for effective design of tunnel supports. The rock mass met with at the test locations at Additional Surge Gallery was very good quality Compact Basalt. The depth of overburden in the 9.5 m diameter horse shoe shaped HRT was found to vary considerably ranging between 30 m to 115 m.The rock mass varieties met with at Flat Jack Test locations at extended H.R.T. were mainly Porphyritic as well as massive and jointed varieties of Compact Basalt and very soft to moderately soft varieties of Volcanic Breccia. The Flat Jack Method (BIS 13946 (Part IV),1994), was employed to measure the deformability of rockmass as well as in-situ stresses at 4 locations at the Additional Surge Gallery and 70 locations at the extended Head Race Tunnel.

The method consisted of cutting a thin slot in the rock surface by drilling a series of overlapping holes to relieve the rock of the original stresses. Because of the stress relief, sides of a slot converged. The convergence of the slot was measured between two reference points fixed at a known distance on either sides of the slot prior to cutting of the slot. The Flat Jack was then embedded tightly in the slot and later pressurized by means of hydraulic pump to neutralize the convergence. This pressure at neutralization, termed as cancellation pressure was then corrected to account for the influence of the slot size, reference pins position and the stress acting parallel to the major axis of the slot, so as to obtain induced stresses acting perpendicular to the slot.

A number of loading and unloading cycles at each test location resulted in computing modulus of deformability of in-situ rock mass as well as plotting of stress-deformation envelopes. After knowing the induced stresses, the stress concentration factors at the tested elevations based on the excavated shape at the section and unit loading in vertical and horizontal directions respectively were obtained by using 2D Stress Analysis Software package Examine 2D and in turn in-situ Vertical Stress (sV) and Horizontal Stress (sH) were evaluated. On basis of these results, the lining design for the entire HRT was done by the project authority. Based on rock mass varieties at different sections as well as amount of available rock cover, various types of supports comprising of rockbolts, shotcrete, PCC and RCC linings of thicknesses varying from 40 cm to 55 cm were adopted.

Ghatghar Pumped Storage Scheme comprises Upper and Lower Reservoirs of gross storage 5.87 Million-cum and 3.87 Million-cum respectively. The upper reservoir was created by constructing 14.5 m high and 478 m long upper dam and two saddles. The lower reservoir is created by constructing 84 m high and 415 m long lower dam. The two reservoirs have a head difference of 420 metres. Both the dams are roller compacted concrete (RCC) dams. RCC technique is being adopted for first time in India for dam construction. The studies were conducted at CWPRS to determine thermai properties of the RCC mixes and estimation of suitable placement temperature. was delayed by increasing dosage of admixture (1% by weight of cement + fly ash). Delaying setting time lowers rate of gain of strength and rate of adiabatic temperature rise. To accommodate the effect of retarder, the temperature monitoring of the RCC mixes had to be continued for a period of 30 to 40 days.

The adiabatic calorimeter developed at CWPRS was used for the studies. Its temperature controller was modified for measuring and controlling small change in rate of temperature rise. The calorimeter was further insulated by additional insulation to avoid heat loss during the course of 30-40 days of measurement of temperature. The effect of long term measurement, addition of fly ash and retarder were evaluated. The studies helped in estimating the effect of fly ash and retarder in reduction and rate of temperature rise. The results of the tests were used as input parameters to compute the maximum temperature rise, suitable placement temperature and pre-cooling requirement. In house developed computer program was used for analysis.

Initially thermal properties of the three concrete mixes with different cement to fly ash ratio were evaluated.

On the basis of these studies, a final RCC mix with cement to fly ash ratio of 88:132 was adopted for use in construction of dams. Thus in the final mix 60% of cement component was replaced by fly ash. The properties of final RCC mix were then used to compute the suitable placement temperature for 0.3 m lift height and 24 hours lift interval. For ease of rolling of concrete and for ensuring adequate bonding between layers of RCC, the setting time of RCC mix Janai Shirsai Lift Irrigation Scheme is located in Daund Taluka near Patas in Maharashtra. The intake sump receives water from Khadakwasla Irrigation Canal.

Janai project has three stage pumping to irrigate land admeasuring 13,835 hectares. The Maharashtra Krishna Valley Development Corporation (MKVDC) has executed the Scheme and it is in the final stage of completion. The subject scheme uses 7 Vertical Turbine (VT) pumpsets rated for 525 kW power, 0.920 m^3/sec discharge and 48.9 m head for Janai Stage I. Similarly, for Janai Stage II, there are 7 VT pump sets rated for 658 kW power, 0.820 m^3/sec discharge and 69.77 m head and a third set of 8 VT pumpsets rated for 446 kW power, 0.495 m^3/sec discharge and 78.06 m head at Shirsai Pump house. The contractual clause for the said installations stipulated that the contractors should carryout performance guarantee tests as per provisions of IS 9137 in presence of a third party expert agency. CWPRS participated in the tests conducted on all the 22 pump sets, as the expert third party neutral agency calibrated the ultrasonic flowmeter which was used during tests. The site tests were conducted with duly calibrated instruments such as tachometer, thermometer, vibration meter, voltmeter, ammeter, frequency meter and wattmeters, etc. to verify the performance behaviour of the pumps in terms of the following:

Pump head and efficiency against discharge Motor power variation against discharge Comparison of the results obtained against those guaranteed by the contractors

During the course of tests, it was observed that sizeable quantity of silt had deposited around the suction pipes in the intake well, resulting in unfavourable hydraulic conditions on the suction end of the pumpsets. The project authorities on recommendations of CWPRS made elaborate arrangements to get the sump cleaned thoroughly at Shirsai pump house. This provided better flow conditions on the suction side for the pump tests.

THIRD PARTY PERFORMANCE GUARANTEE PUMP TESTS FOR JANAI (I&II) & SHIRSAI LIFT IRRIGATION SCHEME, MAHARASHTRA

It was recommended that this aspect be taken care of during routine operation of these pumpsets in future. Head developed by the pump was measured by digital/Bourdon pressure gauge calibrated against dead weight tester. Discharge was measured by deploying ultrasonic flowmeter. Hydraulic power delivered by the pump was found out from the product of head and discharge. Electrical power input to the motor was measured through the wattmeters. Overall efficiency was calculated from the ratio of hydraulic power delivered to the electrical power input. Bowl efficiency was computed after making due allowances for losses in bearings and line shafts.

The data recorded during the tests was at available mains frequency which was monitored from the instruments fitted on the control panels on these pump houses. The observed results were scaled to 50 Hz. It was ensured that all the measuring instruments, selected for conducting performance guarantee tests were having better accuracy than the limits specified in IS 9137. The discharge measuring section was selected at a distance of 5 times the pipe diameter (D) from the pump delivery flange and the head measuring section was located at 3 D from delivery flange. This ensured that the required velocity profile at these sections is fully developed for accurate measurements. During the tests the discharge was varied using the throttle valves on the delivery side and every time readings of discharge, head, power, speed, mains frequency, current, voltage etc. were taken. Thereafter the pump performance characteristics in terms of head-discharge, power-discharge and efficiency-discharge were plotted. Crucial performance parameters like efficiency at duty point, guarantee factor-to show how closely the pump met the specified duty conditions, power at duty point etc. were calculated and the findings were furnished to the Project Authorities.

All the pumps tested were found to have satisfactory and consistent head-discharge characteristic All the pumps tested met the condition of having guarantee factor greater than 1, which indicated that the pumps satisfactorily met the specified head and discharge All the pumps met the guaranteed efficiency The tests helped MKVDC to ensure that the supplied pumps did meet the stipulated performance specifications.

NUCLEAR DENSITY LOGGING FOR MEASUREMENT OF IN SITU BULK DENSITY OF MASONRY AT DUDHGANGA DAM, MAHARASHTRA

The 73 m high Dudhganga Masonry dam was constructed in the year 1989, across river Dudhganga, a tributary of river

Krishna in Kolhapur district of Maharashtra. The main purpose of the dam is for irrigation but it also has a powerhouse to generate electricity of 24 MW. Excessive leakage was noticed in the drainage gallery and it increased with the rise in reservoir level. Maximum total leakage measured in the drainage gallery as well as from the downstream face of the dam was about 350 lps. In view of the excessive leakage from the dam, an expert committee was set up to study safety of the dam which suggested the measurement of density of masonry. Accordingly, the CWPRS team carried out Nuclear logging in the three Nx size boreholes drilled in the masonry to depths of 26 m to 40 m in monoliths 5, 6 & 7. Caliper logging was also carried out for identifying gaps / voids present in the masonry. The results of the studies indicated that the bulk density of masonry in general varied from 2.27 g/cc to 2.52 g/cc. The density of masonry below 2.3 g/cc was attributed to the presence of gaps / voids created by leaching of cement mortar material due to excessive leakage. It was recommended to undertake tracer studies to delineate the path of seepage through the body of the dam.

RAW WATER INTAKE FOR BHUSHAN STEEL AND STRIPS LIMITED ON RIVER BRAHMNI, ORISSA

Bhushan Steel & Strips Ltd. (BSSL) a leading producer of cold rolled sheets and galvanized sheets had proposed to set up large integrated steel plant at Meramandali town in Angul district of Orissa at about 120 km North West of Bhubaneshwar. A 600 MW coal fired captive power plant is also a part of the project. The total requirement of water has been estimated to be about 2.89 m^3/s which would be drawn from river Brahmni at about 5 km from plant site. Studies were conducted at CWPRS for deciding the appropriate location and hydraulic design of intake for assured water supply with minimum operational and maintenance problems.

The satellite imageries, toposheets and river cross-sections were studied and the analysis of available prototype data was carried out for finding minimum flow, minimum water level, maximum flow and corresponding HFL, bed and bank material properties. From the study of satellite imageries of different years, it was observed that the Mangalpur and Kharagprasad ghat are the two locations along Brahmani river where the deep channel is always close to the right bank. The right bank near Mangalpur or Kharagprasad ghat was found suitable. However, the location near Kharagprasad ghat has advantages that the river flows in a single channel and the deep channel is hugging the right bank.

The Kharagprasad Ghat is being used as a ferry point over the years as seen from 1930 toposheets. Moreover, the right bank is relatively stable and the local inquiry supports the fact that the deep channel has never left this location. Therefore, the intake location at about 130m downstream of Kharagprasad ghat was recommended for BSSL plant.

From the analysis of flow data and measurements during the site visit in May 2004, it was estimated that the water level corresponding to minimum flow of 40 m^3/s gives a depth of flow of 1.50 m in the deep channel. A 30 m long weir along the right bank with crest at about 0.50 m above river bed followed by a silt trap and a pump-sump have been recommended. Considering the minimum depth of flow of 1.50 m available in deep channel, the sill level of pump-sump has been proposed at 2 m below minimum water level so as to satisfy submergence requirement. Necessary bank and toe protection in the form of stones in crates and synthetic filter below were also recommended for a distance of 50m upstream and 20 m downstream of the intake.

INTAKE WELL OF PAPER MILL OF J.K. CORPORATION LTD. IN RIVER NAGAVELI AT JAKAYPUR, ORISSA

The paper mill of J.K. Corporation Ltd. is situated in Rayagada district of Orissa State. An intake well for drawing water for the mill is located in river Nagaveli near right bank. The intake well was commissioned in 1962 to draw raw water of 10 mgd for 18000 TPA capacity. The functioning of intake well with regard to drawl and availability of water was satisfactory till 1982. The Singpur nalla joins at about 250m upstream of existing intake with its entry facing upstream of Nagaveli river. This nalla experiences flash flood with very high sediment load which gets deposited in the river near the right bank and therefore the flow gets deflected towards the left causing erosion of the left bank. The erosion continued in a reach of about 300m. Due to morphological changes in the river over the years, the major flow during lean season concentrated on left side and the right bank where the intake was located, remained dry and the Project Authorities at the paper mill of J.K. Corporation Ltd. were facing problem of non-availability of water at intake. Various temporary measures such as construction of stone barrier, sand bund along the left bank were being adopted by Project Authorities to meet required quantity of water. Further, studies were undertaken at CWPRS to suggest remedial measures for ensuring the availability of required quantity of water at intake particularly during lean season. As a part of the studies, CWPRS Officers inspected the intake site and detailed desk studies were carried out subsequently.

During the site inspection, Project Authorities informed that heavy silting was experienced at the intake during floods. Frequent dredging was also necessary during lean season for maintaining required depth of water. Based on site inspection, CWPRS suggested short term and long-term solutions for the existing problems. A short-term measure of construction of porcupines in the reach of active erosion on the left bank was suggested based on site inspection, data made available and past experience. It was indicated by the Project Authorities that part implementation of short-term measure resulted in improvement of flow conditions near intake. Various long-term measures were examined with the available data. These included pilot channel, construction of weir, realignment of entry of Singpur nalla with Nagaveli river, diversion of Singpur nalla downstream of intake and re-location of the intake at the nodal point upstream of nalla. Considering the merits and demerits of these alternatives, it was considered that locating intake in the river upstream of the confluence of Singpur nalla in the bend along the outer curve would be a permanent long-term solution. Re-locating the intake would ensure the availability of required quantity of water even during lean season and would not pose any serious operational and maintenance problems as were being experienced presently.

However, an open intake on the bank with low-level weir at the entry followed by silt trap would be a better alternative from the point of view of siltation problems. It was recommended that until the long-term measures were implemented at the proposed site, short-term measures of streamlining of entry of Singpur nalla along with provision of porcupines on the left bank be continued and river behavior in the reach be monitored.

NERADI BARRAGE ON RIVER VAMSADHRARA, ANDHRA PRADESH

The river Vamsadhara flows through the lower reaches of the states of Andhra Pradesh and Orissa. A Barrage is proposed to be constructed across this river at Neradi village of Srikakulam District, Andhra Pradesh. Canals are also proposed on both the banks of the river to cater for irrigation. The barrage would consist of total 36 bays of which 30 numbers would be spillway bays having a clear width of 18.29 m. The four number of under sluice bays on right bank and 2 under sluice bays on left bank having a clear width of 12.19 m were also proposed.

The studies were conducted at CWPRS for estimation of backwater effects and examination of sufficiency of upstream protection arrangements by using physical model (scales-1/200 H, 1/40 V) and 1-D mathematical model HEC-6. The studies were carried out for four discharges 1000 m^3/s, 2832 m^3/s, 5663 m^3/s and observed maximum 16990 m^3/s. The studies revealed for discharge equivalent to 16,990 m^3/s that: Back water length after the construction of Neradi Barrage would be of the order of 6 km upstream of barrage Increase in water level at cross section No.2 i.e., 2 km upstream of barrage over the existing condition was observed as 0.72 m The rise in water level at cross section No.6 i.e. 6 km was of the order of 0.16 m. However, the rise in water level diminished rapidly beyond this cross section. With barrage in position, the maximum velocity was observed as 6.17 m/s along the cross section of barrage. Upstream protection measures i.e., concrete blocks and loose boulder protection were therefore, designed accordingly Left and right guide bunds were designed as per BIS Code of Practice to ensure streamline flow through barrage

SHEAR STRENGTH PARAMETERS OF FOUNDATION ROCKMASS FOR UPPER TUNGA PROJECT, KARNATAKA

A 26.2 m high and 770 m long composite dam is under construction across river Tunga at about 100m downstream of the existing Anicut at Gajanur village of Shimoga district, Karnataka. The dam consists of 15.20 m high and 117.50 m long earthen dam with central puddle core at the left flank, a Power block of 52 m length to generate 19.5 MW power, non-overflow sections of 18.50 m length on the left flank and of 126 m length on the right flank. A 321.50 m long high co-efficient weir type concrete Spillway is at the central portion with 22 numbers of radial crest gates of size 11.75 m x 4.74 m to discharge design flood of 7362.40 m^3/s. The field and laboratory studies were conducted by CWPRS to determine the in-situ shear strength parameters and other engineering properties for foundation-rock interface at the downstream of Spillway block nos. 14,18 and 20.

These data would be utilized as input data for estimation of dam safety since the foundation rock mass is schist.

The general rock mass forming the dam foundation consists of good quality granites with occasional schistose zone. The foundation rockmass, exposed as outcrop was found to be fresh and hard rock of Schistose variety. In-situ shear tests were carried out on six concrete blocks of sizes 60cm x 70cm casted on foundation rock. For each block, the test procedure involved applying a pre-determined normal stress on the test block and then applying shear stress in small increments till the failure of the block. All six blocks were tested under different normal load conditions to evaluate the shear strength parameters for the dam-foundation interface. The study of failure surface from the upturned views of the blocks revealed that out of six blocks, only two blocks failed along the contact whereas the others failed mostly through rocks and partly through contact. The estimated values of shear strength parameters of the rock blocks i.e. cohesion (c) and friction angle (f) were found to be 10 kg/cm^2 and 59° respectively.

Laboratory studies were conducted on four rock core samples collected from project site representing the foundation rock mass to determine various other engineering properties. The values of properties such as Density, Specific Gravity, % Water Absorption, Static Modulus of Elasticity, Poisson's ratio,

Unconfined Compressive Strength and Hardness were found to be varying between 2.75 to 2.85 gm/cc, 2.81 to 2.93, 0.10 to

0.43, 6.58 x 10^5 kg/cm^2 to 8.90x 10^5 kg/cm^2, 0.16 to 0.24, 215 kg/cm^2 to 632 kg/cm^2 and 14 to 33 respectively.

STUDIES FOR UJH RIVER LEVEL CROSSING, JAMMU & KASHMIR

The Ravi canal project in Jammu and Kashmir is the largest irrigation scheme in the Jammu area. The Ravi canal takes-off from the right bank of the river just downstream of the tailrace of powerhouse of the Thein dam project. The 82 km Ravi canal enroute crosses a number of major, medium and minor streams including Ujh river. Ujh river originates from high altitude hills in Tehsil Billawar and Ramnagar of district Udampur and Kathua in J&K state. The Ujh river is a sub-mountainous river which is shallow, wide and has a braided channel pattern. The river is having steep slopes of the order of about 1:150 at the level crossing with riverbed consists mostly of shingles and boulders.

The Ujh barrage constructed at level crossing was commissioned in 1980 with a waterway of about 356 m for augmenting of Ravi canal flow. The existing Ujh river barrage level crossing has 18 bays of 16m span with pier thickness of 1.75m for spillway part and 3 spans of 8m, 8m and 16m for undersluice portion. The barrage was designed for a discharge of 5000 m^3/s. In 1987, a high flood of 8835 m^3/s was experienced which resulted in extensive damage to barrage and its appurtenant works.

The design discharge was therefore revised by Central Water Commission to 8400 m^3/s. Since the existing waterway of 356 m for the barrage was inadequate to pass the revised design discharge of 8400 m^3/s, it was proposed to extend the waterway of barrage by about 322 m on the left side. Model studies (scales: 1/200 H, 1/70V) were conducted at CWPRS to study various hydraulic issues such as adequacy of waterway, protection works, flow conditions, energy dissipation arrangements etc. with proposed extension of barrage waterway.

The model studies were carried out with extended waterway and re-laid bed as per the fresh survey. It was expected that with proposed extension of waterway of 322 m on the left side, the overall flow conditions and discharge distribution would improve for the revised design discharge. However, the studies indicated that due to cross slope and steep longitudinal slope, the concentration of flow was towards right at the undersluice and adjoining spillway portion and discharge distribution at the revised design discharge was seen to be 75% and 25% through existing and extended waterway respectively.

It was therefore, concluded that there was no significant improvement in the overall flow conditions and discharge distribution with the proposed extension of waterway. From the studies, it was inferred that there is a scope for reduction in the total waterway and further studies would be required for optimization in the length of proposed extension of waterway.

Under the Choudhari Charansingh Pariyojana, the Uttar Pradesh Irrigation Department (UIPD) undertook procurement and installation of large quantities of submersible pumps to tap the ground water potential for purposes of irrigation.

The procurement is done on the basis of competitive bidding. One of the important clauses included in the tender enquiry while inviting bids was to get sample designs of these submersible pump sets tested as per testing procedure contained in IS 9137 for their hydraulic performance as well as the performance of their associated motors at CWPRS. Such tests would enable UPID to verify the specific performance guarantees quoted by the bidders in their offer against those achieved in the actual tests conducted by a neutral laboratory.

CWPRS conducted following performance tests on ten samples of these pumpsets rated for 21 m and 42 m head with 100 cum/hr discharge. Determination of hydraulic performance characteristics viz. variation of head, power input and overall efficiency against discharge covering a minimum range of + 10 % to-25 % of rated head from guaranteed duty point for the 21m head and 42 m head pumpsets Conducting over voltage and under voltage tests at 456 volts and 353 volts respectively Determination of temperature rise of the submersible motor, when the pump is overloaded to 32 % of its rated capacity Analysis of the experimental results by scaling the pump motor performance to 50 Hz mains frequency and to ascertain the performance of these sample pumpsets against guaranteed values

HYDRAULIC PERFORMANCE AND OVERLOAD TESTS FOR SUBMERSIBLE PUMP SETS

All the tests were carried out in accordance with IS 9137 but uncertainty level in the measurements carried out during the tests was much better than IS stipulations, especially for the flow rate and pressure measurement.

It was recommended that: The submersible pump set having highest efficiency and lowest temperature rise should be preferred as this would ultimately result in energy efficient operation and long-term reliability of the pump set.

While selecting the pump sets on technical merits, preference be given to the submersible pump unit having highest guarantee factor, reason for this being the higher the guarantee factor, closer it would operate to the guaranteed duty requirement.

It would be worthwhile to randomly select pump sets from the actual batch of supply and get these tested at CWPRS for ensuring better quality of performance of the actually supplied pumpset. The Tala H.E. Project, Bhutan envisages construction of a 91 m high concrete gravity dam across river Wangchu near Honka and an underground power house near Tala having installed capacity of 1020 MW (6 units of 170 MW each). The spillway in the middle portion of the dam comprises five sluices of size 6.5 m wide x 13.15 m high with invert at EL.

1320 m and one span of overflow portion of size 4 m wide x 3 m high with ogee shaped crest at EL. 1360 m. All the sluices and the overflow portion have been equipped with radial gates for surplussing flood water. Ski-jump bucket has been provided for energy dissipation for both overflow portion and sluices. The spillway capacity has been provided to pass standard project flood (SPF) with peak outflow of 8575 m^3/s at full reservoir level (FRL) EL. 1363 m and the maximum probable flood (PMF) with peak outflow of 10,600 m^3/s at reservoir water level EL. 1365 m. Earlier, the studies were conducted at CWPRS using the 1:60 scale comprehensive physical model and considerable modifications were suggested for spillway and stilling basin configurations.

The studies on the original design of sluices with elliptic type of roof indicated that the sluices would be subjected to large negative pressure for discharges higher than 9000 m^3/s

passed under ungated operation of spillway and might lead to cavitation damage. Studies were therefore, conducted for minimizing the sluice roof pressures by changing the sluice profiles and for evolving discharging capacity curves for partial gate operation of sluice spillway on a separate 1:40 scale 2-D sectional model. As the Tala H.E. Project is located in the highly seismic Himalayan tectonic province, the seismological studies were also carried out by CWPRS. The studies were conducted with ungated operation of spillway sluice for maximum probable outflow flood of 10600 m^3/s, SPF of 8875 m^3/s and 75% of SPF i.e. 6431 m^3/s. The pressures over the roof profile of sluice for lower discharges with gated and ungated operation were acceptable. Three alternative roof profiles were studied. For alternative-III with circular (45°) profile, the negative pressures were comparatively less than all other alternatives designs i.e. cubic and circular (37.8°). The cavitation index corresponding to maximum negative pressure was greater than the critical value of 0.2. The Alternative III profile was recommended. The discharging capacity curves for spillway with all sluices operating under full and partial gate operation were suggested. It was observed that under fully open condition of gates, a discharge of 10200 m^3/s would be passed at reservoir water level EL. 1363 m.

On the basis of historical seismicity, various tectonic features in the proximity of Tala project were indicated to be seismically active. To delineate these features and to assess their seismogenic potential, comprehensive microearthquake studies were taken up by employing one microearthquake (MEQ) recorder and strong motion

HYDRAULIC AND SEISMOLOGICAL STUDIES FOR TALA H.E. PROJECT, BHUTAN

Accelerograph (SMA) at Tshimalakha site which is close to dam and other at Piping observatory, at about 30 km d/s of dam. The data were recorded during the period January 2002 to December 2003 and these microearthquake records were analysed to estimate the epicentral distances from the recording station and the magnitudes of the microearthquakes.

2

Identification of Molecular Receptive Ranges

INTRODUCTION

The sense of smell is crucial for most animal species. It is critical for food-finding, reproductive behaviour, predator-prey relationship, kin and mother-infant recognition, homing behaviour and nest finding. The importance of the olfactory systems is reflected in the proportion of the genome that is devoted to the olfactory receptor proteins, e.g. comprising 3-5% in human and mouse (Young and Trask, 2002; Zhang and Firestein, 2002). For a long time the human sense of smell was considered as the most enigmatic of our senses. An intriguing question was; what mechanism could explain our ability of recognizing and remembering more than 10 000 distinct odorants (Buck, 2004). Buck and Axel (1991) performed a breakthrough by the discovery of the large family of olfactory genes in the rat. Buck and Axel were in 2004 honoured with the Nobel Prize in physiology and medicine for this study and the following series of pioneering work on the subject. The knowledge about the olfactory genes is obviously important for studies of the function of the olfactory receptor neurons (RNs), both in solving the transduction mechanisms and the specificity of the RNs. In the search for which odorants the olfactory receptors are evolved, insects are suitable model organisms.

Herbivore species are particularly interesting, since many of them share the same plant species and their survival depends on olfactory cues in locating their host for feeding and reproduction (mating and oviposition).

The Insect Olfactory System; Anatomy of the Olfactory Pathway

The numerous olfactory organs in insects, the sensilla, are mainly located on the antenna (Schneider and Steinbrecht, 1968). The lepidopteran antenna consists of two proximal segments, the scape and pedicel, and the long flagellum. In heliothine moths, the flagellum consists of ~ 80 annuli that carry numerous sensilla mediating information about different modalities, including chemo-, mecano-temperature-, and humidity sensation (Almaas and Mustaparta, 1990, 1991; Jørgensen, 2003; Kvello, 2003; Lassa, 2004). Like in other Lepidopterans, the olfactory sensilla of the moth antenna outnumber by far those of other modalities. The general classification into various morphological types, like sensilla trichodea, s. basiconica, s. auricillia and s. coeloconica also applies to olfactory sensilla of heliothine moths (Jefferson et al., 1970; Steinbrecht, 1973; Hallberg, 1981; Keil and Steinbrecht, 1984; Almaas and Mustaparta, 1990; Almaas et al., 1991; Koh et al., 1995; Færavaag, 1999).

Extensive studies have been carried out on the structure of s. trichodea and s. basiconica, involved in pheromone and plant odour detection, respectively (review Steinbrecht, 1997). The cuticle wall of these sensilla is perforated by pores allowing the air-borne volatiles to enter the lumen, which is filled with receptor lymph surrounding the dendrites of the RNs. The membrane spanned receptor proteins are located in the dendrite of these bipolar sensory neurons.

The axons of the antennal RNs form the antennal nerve and project directly to the deutocerebrum, the first relay station of the antennal sensory pathway (Homberg et al., 1989). The bilateral deutocerebrum consist of two distinct regions called the antennal lobe (AL) and the antennal mechanosensory and motor centre (AMMC, also called the dorsal lobe in other species). The olfactory RNs send their axons into the AL, whereas the AMMC receives axons from the mechanosensory neurons (Homberg et al., 1989). Synapses between the RNs and antennal lobe neurons are located in numerous glomerular structures of the antennal lobe. These structures are functional units and represent a physical basis for mapping odour qualities. In herbivorous Lepidopterans, many studies have shown a separation of the glomeruli involved in the two systems of pathways mediating pheromone information and plant odour information.

In species of Heliothinae, the three male specific glomeruli constitute the macroglomerular complex (MGC) dedicated to the pheromone information, and 60-62 ordinary glomeruli dedicated to plant odour information (review Mustaparta, 2002). In the AL two major morphological types of neurons receive and process the olfactory information from the antennal sensory neurons. The local interneurons with arborisation in many glomeruli mediate information within the antennal lobe, whereas projection neurons branching in one or a few glomeruli have an axon conveying information out of the AL to higher order neurons in the protocerebrum. These are located in two areas, the mushroom bodies, shown to be important in learning and memory of odours (review Menzel, 1999), and the lateral horn which is a pre-motoric area (Strausfeld, 1976). In moths, including heliothine, the axons of the projection neurons follow three major tracts from the AL to the protocerebrum, the inner-the outer-and the middle antenno-cerebral tract (Homberg et al., 1988; Rø et al., 2003).

Peripheral Events

The binding of and interaction between the odorant and the receptor proteins leads to an intracellular cascade reaction (the transduction events), which results in opening of ion channels and depolarisation of the membrane. Lancet and Pace (1987) was the first to identify G-proteins in the olfactory epithelium of vertebrates, suggesting that activation of this protein by the odorant-receptor interaction is the first step of the cascade. In insects, the presence of G-proteins in olfactory neurons was demonstrated by Breer and co-authors (1988). The cascade leading to production of IP3 (inositol 1,4,5-trisphosphate) as second messenger is, in insects, considered to be the major excitatory pathway, opening the cation channels (Breer et al., 1990; Wegener et al., 1993; Stengl, 1994). In vertebrates, cAMP (adenosine 3,5-monophosphate) is the second messenger for excitation (Nakamura and Gold, 1987; Breer et al., 1990; Breer, 2003b). cAMP has also been indicated as a possible second messenger in insects (Krieger et al., 1999). The odorants reach the receptors via odorant-binding proteins (OBP) present in the receptor lymph.

Two major groups of binding proteins are classified in insects, the general odorant binding proteins (GOBPs) and the pheromone specific proteins (PBPs), each consisting of several sub types (Steinbrecht et al., 1992; Laue et al., 1994; Zhang et al., 2001). OBPs are assumed to function as transporters of the air born (hydrophobic) volatiles that have to pass through the liquid receptor lymph in order to reach the receptor proteins. Since OBPs show selective binding to some odorants they may also serve as a filter, protecting the receptors from being exposed to all kinds of volatile compounds. It has been questioned whether OBPs also contribute in odorant-receptor binding and in inactivation of the odorant-receptor complex (Prestwich et al., 1995; Steinbrecht, 1998; Kaissling, 1998; Mohl et al., 2002; Pophof, 2004). Another hypothesis is that the OBPs release the odorants close to the dendrite membrane due to conformational changes caused by the charged membrane (Wojtasek and Leal, 1999). Since the OBPs are present in the chemosensory systems of terrestrial vertebrates and insects, it is suggested that these proteins may be a molecular adaptation to terrestrial life (Breer, 2003a).

"The Logic of the Sense of Smell"

Olfactory receptor genes first identified in the rat by Buck and Axel (1991) is one of the largest known mammalian gene families, in rats and mouse comprising nearly 1000 genes expressed exclusively in the olfactory tissue. In the many molecular biologically studies that followed, the general finding was that the olfactory information is handled by a large and species-specific number of receptor proteins (Buck and Axel, 1991; reviews Mombaerts, 1999, 2004, Keller and Vosshall, 2003, Breer, 2003a). Studies conducted over the past decade have shown that one type of olfactory receptor gene is expressed in a given subset of RNs (Ressler et al., 1993; Vassar et al., 1993; Clyne et al., 1999; Vosshall et al., 1999; Hallem et al., 2004). Candidate receptor proteins have also been identified in Heliothis virescens, showing expression of only one type in each neuron (Krieger et al., 2002, 2004).

The olfactory receptor proteins show low homology across phyla. Only one subtype sharing a high degree of sequence identity in several species is co-expressed with other receptor proteins (Clyne et al., 1999; Fox et al., 2002; Krieger et al., 2003; Breer, 2003a). It is assumed that this particular protein has a role other than odorant recognition (Breer, 2003a). Furthermore, molecular biological studies of both vertebrates and invertebrates have shown that each subsets of RNs, expressing the same type of receptor proteins, projects in one or two specific glomeruli of the primary olfactory centres (the antennal lobe in insects and the olfactory bulb in vertebrates) (Axel, 1995; Treloar et al., 2002; Keller and Vosshall, 2003; Mombaerts, 2004). This principle, called "the logic of the sense of smell" suggests a certain relationship between the number of RN types and the number of glomeruli in the primary olfactory centres (Axel, 1995). In insects, this principle has been demonstrated in Drosophila, showing that each subset of RNs projects exclusively in one (or sometimes two) homologous glomeruli in each antennal lobe (Gao et al., 2000; Vosshall et al., 2000; Keller and Vosshall, 2003).

Numerous electrophysiological studies have been performed with the aim to functionally classify olfactory RNs (among others, Sicard and Holley, 1984; Ma and Shepherd, 2000; Duchamp-Viret et al., 1999; reviews Shepherd, 1984, Masson and Mustaparta, 1990, Hildebrand and Shepherd, 1997, Todd and Baker, 1999, Mustaparta, 2002, Korsching, 2002). These studies of both vertebrates and invertebrates have shown a large variation of the molecular receptive ranges, from RNs being narrowly tuned and falling into distinct types, to broadly tuned neurons, often with individually different molecular receptive ranges. The pheromone olfactory receptors in heliothine moth are particular well studied. Since the first identifications of the female produced sexual pheromones in these species, listed in Arn et al. (1992), these biologically important odorants and other interesting chemical analogues have been available for detailed studies of the RN specificity (Masson and Mustaparta, 1990, Mustaparta 1997).

Electrophysiological studies have demonstrated a functional classification of three or four RN-types tuned to the insect produced signals in each species. These RNs are characterised by a narrow tuning to one compound and considerably weaker responses to a few chemical analogues. Furthermore, functional tracing of single RNs have demonstrated that the axon terminals of each RN type project in one of the three or four glomeruli of the male MGC (Hansson et al., 1995; Berg, 1998; Berg et al., 1998). These findings have also been supported by optical recordings using Ca2+ imaging (Galizia et al., 2000). Thus, the results from studies of pheromone receptors in heliothine moth correlate well with the principle of one subset of RNs projecting in one glomerulus.

Chemical Aspects of Insect-plant Interaction

Plants produce hundreds of compounds that are important in their interaction with insects and other organisms. These compounds are termed secondary metabolites, whereas those essential for growth and development of the plant are called the primary metabolites (Hartmann, 1996). Traditionally, secondary plant metabolites like the volatile compounds emitted from flowers and leaves were looked upon as by-products with no relevance. Ehrlich and Raven (1964) were among the first to suggest that plant produced secondary metabolites are evolved in a co-evolutionary arms race of plant defences and herbivore responses. The plants produce and release volatiles, e.g. for attracting pollinators. Pollinators might as well be herbivorous using these signals for host location (Harborne, 1993). After pollination plants are able to turn off the advertisement to pollinators by gene down regulation, which makes the plant less exposed to herbivory (Tollsten and Bergström, 1989; Dudareva and Pichersky, 2000). Plants may obtain a competitive advantage by producing other specific and reliable chemical signals that repel putative herbivores (direct defence) or attract natural enemies of the herbivores (indirect defence) (e.g. Bernays and Chapman, 1994; Schoonhoven et al., 1998).

Particularly the flowers and seeds, the reproductive parts of a plant, important for the plant fitness, need to be defended against herbivory. The heliothine larvae, mainly feeding on the growing and reproductive parts of the plants, are hazardous to the host plants (Fitt 1989). As defence, the plants produce and accumulate toxins that are damaging to the insects. As a reciprocal response, the insects may detoxify or excrete the toxins. Generalist feeders, which are exposed to a wide spectrum of toxic compounds produced by the plant defence systems, have a welldeveloped detoxification system, exemplified by the high activity of the MFO-enzyme system (mixed function oxidises) (Brattsten, 1983). Being toxic also to the plant themselves, these compounds are often produced as pro-toxins and are constitutively accumulated in special organs like vacuoles and glandular trichomes (Hartmann, 1985).

Plants are continuously interacting with their surroundings. The profile of volatiles varies during exposure of many biotic and abiotic factors, like nutrition access (e.g. nitrogen deficits), microbial infestation, exposure to UV light and ozone, high temperatures or auto-oxidation by the surrounding air (Janssens et al., 1992; Pichersky and Gershenzon, 2002; De Moraes et al., 2004). The profile of emitted volatiles also shows diurnal and seasonal variations (Hedin, 1976; Dudareva et al., 1999; Kolosova et al., 2001). All these factors might influence the signals exploited by herbivores in their host location. This tremendous complexity and variability of plant volatiles is very challenging in the investigation of biologically significant odorants used by insects and other organisms.

Many studies have been performed on tritrophic interactions, i.e. between plants, herbivores and herbivorous predators or parasitoids. Particularly interesting are the findings showing increased production and release of volatiles during caterpillar attack (Turlings and Benrey, 1998; Paré and Tumlinson, 1999; Dicke and Van Loon, 2000; Schmelz et al., 2003). Furthermore, profiles of compounds systemically induced during herbivory show species specificity, as regards quality and quantity, which is also shown for attack by heliothine species (Mori et al., 2001; Röse and Tumlinson, 2004; De Moraes and Mescher, 2004).

Thus, the volatiles released by plants in response to insect feeding are directly associated with the feeding herbivore species. This induction is caused by activation of a series of genes that up-regulate the specific defence in plants (Halitschke and Baldwin, 2003; Frey et al., 2004). The larval oral secretion contains several factors (e.g. volicitin) that induce plant defence responses (Alborn et al., 1997; Mori et al., 2001; Spiteller et al.,

2001). Various toxins, like the tannins and gossypol, present in high amounts in flower buds of cotton (Gossypium hirsutum) are enzymatically ignited or induced in response to caterpillar feeding (Bezemer et al., 2004). These toxins have a negative effect on the development and survival of several cotton pest insects (Sharma and Agarwal, 1982; Stipanovic et al., 1990; Hedin et al., 1991). In Nicotiana species, the content of nicotine increases after herbivory or mechanical damage (Euler and Baldwin, 1996). These toxic plant metabolites are deterrents (inhibit feeding) to several pest insects and protect plants against predation (Bernays and Chapman, 1994).

The complex blends of volatiles produced by a plant can be trapped by various methods of headspace collection, distillation or extraction (review Silverstein and Rodin, 1966). More plant constituents present in nature are identified continuously as more sensitive analytical methods are employed. Gas chromatography, which separates different molecules, linked to or followed by mass spectrometry is a common method used for identifying volatile compounds in plants. These compounds belong to many different chemical groups, like short chain alcohols, aldehydes, ketones and esters, aromatic compounds (like benzenoids), mono-and sesquiterpenes (reviews Gibbs, 1974, Smith, 1976, Knudsen et al., 1993, Bernays and Chapman, 1994, McDonough et al., 1994, Ohloff, 1994, Schoonhoven et al., 1998). A few compounds are mainly found in restricted plant taxa, e.g. the isothiocyanates in Brassicacea (reviews Kjær, 1976, Fahey et al., 2001). Others, commonly occurring, are "green leaf volatiles" (mainly sixcarbon alcohols, aldehydes and esters) that are products of the lipid metabolism catalysed by the enzyme lipoxygenase present in green leaves (Hatanaka, 1993; Rosahl, 1996; Croft et al., 1993; Heiden et al., 2003). Some compounds like the terpenoids (linalool, geraniol, limonene, myrcene, E-β-ocimene, farnesene, nerolidol and caryophyllene, among others), are common constituents of flowers, but are also present in vegetative tissues, where they serve as defence compounds (Knudsen et al., 1993).

The Heliothine Moths

The subfamily Heliothinae (Insecta; Lepidoptera; Noctuidae) constitute a large group of herbivore insects, of which the three important agricultural pest species Heliothis virescens, Helicoverpa armigera and Helicoverpa assulta were chosen for the present studies. H. armigera and H. virescens are both generalist feeders (polyphagous) exploiting a wide range of plant species across different families (e.g. Leguminosae, Solanaceae, Malvacea and Compositae) (Fitt 1989; Matthews 1991). Many host plants exploited by one or both species are economically important agricultural crops, like cotton, sunflower, tobacco, maize, chickpeas and sorghums (Zalucki et al. 1986; Fitt, 1989; Firempong and Zalucki, 1990). H. assulta is considered oligophagous, exploiting a more narrow range of plant species, mainly within the family Solanacea (Hill, 1983; Matthews, 1991).

The two genera Helicoverpa and Heliothis are considered monophyletic, i.e. having a common origin (Matthews 1999). For millions of years, the American tobacco budworm moth H. virescens has been geographically separated from the closely related H. armigera, living at the Eurasian, African and Australian continents. The Oriental tobacco budworm H. assulta, partly sympatric with H. armigera, is distributed in Asia and Australia. The species, living at different continents, have been separated for a long time, and presumably exploited different host plant species, at least prior to the introduction of crop hosts they have in common. This might have lead to evolutionary changes of the olfactory system.

The introduction of non-selective insecticides to control pest species, disrupted in many cases the natural balance of herbivore and predator/parasite populations (Bottrell and Adkinsson, 1977). Some insect species, like H. virescens, became new major pests because of their remarkable capability to quickly evolve resistance to the insecticides, which threatens the success of pest control (Fitt, 1989). The increasing awareness concerning the ecologically consequences of the wide-spread use of insecticides enforces the search for ecologically viable alternative methods in pest management programs. Increased knowledge about the sensory receptor system of these species, their behaviour and ecology, may help minimize the level of crop damage as well as the amounts of insecticides used. This is being tested by combining mating disruption by pheromones and precise timing of low level exposures of insecticides. In another attraction-kill strategy, the idea is to use plant odorants to attract the females to a gluing material with insecticides.

Aims of the Thesis

When the study of this thesis was initiated, hardly any work had been carried out on how plant odour information was encoded by the olfactory RNs in heliothine moths. The method of gas chromatography linked to single cell recordings (GC-SCR) was employed and improved for identifying naturally occurring plant odorants that are detected by single RNs and can be considered as biologically relevant. Three species of the subfamily Heliothinae were included in this work, the two polyphagous H. virescens and H. armigera and the oligophagous H. assulta. The American H. virescens is geographically separated from the other two species. H. armigera and H. assulta are partly sympatric in Asia and Australia.

The aims of the thesis elucidated in Papers I-IV were as follows:

1. To identify plant produced volatiles detected by antennal RNs in the three species of the subfamily Heliothinae.
2. To elucidate whether the single RNs can be classified into distinct types according to their specificity.
3. To characterise the plant odour RN types by their molecular receptive ranges, sensitivity and specificity.
4. To compare the specificity of plant odour RN types across the three related species of Heliothinae, with the aim to reveal any differences in the peripheral olfactory system that may have evolved through evolution.

SURVEY OF THE INDIVIDUAL PAPERS

Paper I

The study of paper I was the first carried out in the heliothine moth, with the aim to identify plant odorants by the use of gas chromatography linked to electrophysiological recordings from single receptor cells (GC-SCR). Volatiles released by a large number of host and non-host plants, intact as well as cut materials, were collected by headspace techniques, i.e. by trapping organic molecules from the air surrounding the plants. The volatile constituents were led through a tube containing an adsorbent and were subsequently eluted with a solvent. These headspace mixtures were then used as test samples on the RNs. The gas chromatograph was installed with two columns in parallel, each linked to the electrophysiological setup by a split at the outlet. In this way, half of the effluent is led to the GC detector and the other half out of the oven and into an air stream blowing over the insect antenna.

This made it possible to test each single neuron with the compounds separated via two columns with different properties. The results were obtained as simultaneous recordings of gas chromatograms and neuron activity with responses to the active compounds. A large number of RNs were tested for numerous mixtures of plants volatiles. One particular type of neuron frequently appeared in nearly 80% of the recordings from H. virescens females. The neurons responded with high sensitivity and selectivity to one compound present in several hosts as well as non-host materials. The active compound was identified as a sesquiterpene hydrocarbon by the use of linked gas chromatography-mass spectrometry (GC-MS). Isolation of the compound from a sesquiterpene fraction of cubebe oil provided enough material for identification by NMR (nuclear magnetic resonance). The identification of the compounds as germacrene D was verified by retesting the purified compound via the gas chromatograph, which showed a significant response to germacrene D.

All RNs responding to germacrene D showed a weak response to another sesquiterpene hydrocarbon. However, due to the lack of reference material, this compound could not be identified. Thirteen sesquiterpenes structurally related to germacrene D were found to have no effect. The germacrene D neurons presented in this paper was the first example of a narrowly tuned plant odour receptor type in a polyphagous moth species shown by the use of GC-SCR, where hundreds of naturally occurring plant volatiles were screened on each neuron.

Paper II

In Paper II, the specificity of three other RN types in H. virescens were identified on the basis of results obtained with the methods of GC-SCR and GC-MS. The method with two parallel columns was described in detail in this paper. The advantage by testing the same neurons with the same sample sequentially via a polar and a non-polar column was demonstrated. The various samples collected from host as well as non-host (intact and cut) materials were used in the studies of all four papers included in this thesis. The headspace techniques used for collecting the volatiles was described in paper II. In this study activity of three RNs occurred in the same recordings and these neurons were assumed to be co-located in one sensillum.

Occasionally, one or two of them occurred alone in the recordings, or all three occurred together with a fourth neuron for which the compounds were not identified. By screening the neuron for sensitivity to a large number of plant samples containing hundreds of volatiles, all three neurons were found to have a high sensitivity and selectivity for one odorant (primary odorant) by showing weaker responses to a few other compounds with related structures (secondary odorants). On the basis of the mass spectra of the GC-MS analyses, the primary and secondary odorants were identified for neuron type 1 as E-β-ocimene, β-myrcene, Z-β-ocimene and DMNT (4,8-dimethyl-1,3,7-nonatriene, named homo-myrcene in Paper II), for neuron type 2 as E,E-a-farnesene and E-β-farnesene, and for neuron type 3 as TMTT (4,8,12-trimethyl-1,3,7,11-tridecatetraene, named homo-farnesene in Paper II). The responses by neuron type 1 to E-and Z-β-ocimene and β-myrcene was verified by retesting reference samples. Several of the other primary and secondary odorants were retested in the work of Paper III.

Paper III

The study of paper III, using the same method with two parallel columns linked to electrophysiological recordings from single RNs, showed recordings from females of the three heliothine moths H. virescens, H. armigera and H. assulta. Based on 135 tests by GC-SCR from 52 RNs in the three species, the four co-located RNs reported in paper II were functionally described and compared in the three related moth species. From the study of paper II, the primary and most of the secondary compounds were known for three of the RNs in H. virescens. In paper III, the primary and secondary odorants of the fourth type were identified in H. virescens, and the same four RN types were demonstrated in the other two heliothine species. Additional data on the molecular receptive ranges of the former identified RN types were also provided. Thus the primary (underlined) and secondaıy odorants for the four neuron types in the three related moths were described as follows: For RN type I, E-β-ocimene, β-myrcene, Z-β-ocimene, DMNT and dihydromyrcene, for RN type II E,E-a-, Z,E-a-and E-β-farnesene, for RN type III, TMTT, and for RN type IV geraniol, citronellol, (S)-(+)-and (R)-(-)-linalool, in addition to one unidentified compound.

All neurons of the four types were narrowly tuned, by only responding to these odorants out of hundreds naturally occurring plant volatiles tested. Each RN type of the three species showed similar ranking of primary and secondary compounds according to the response strength, indicating a functional similarity. In addition, all four RN types occurred together in the same recordings of the three species, indicating a similar colocation in the sensilla. These similarities indicate a common evolutionary line of these RNs in the heliothine moths.

Paper III also provides an attempt to trace the axons of the four co-located plant odour RNs into the antennal lobe of the insect brain. The fluorescent dye was applied to the base of the sensillum from which the recordings were made. In one successful staining of H. assulta four selectively stained axons in the antennae and four axon terminals in the antennal lobe were obtained. Three of them were located in different areas close to the entrance of the antennal nerve, and the fourth was located in the ventro-medial part of the lobe.

Paper IV

Using the same methods of GC-SCR (with two parallel GC-columns) and GC-MS, results obtained in the study of paper IV contribute with identification and classification of fourteen out of totally nineteen RN types recorded in the polyphagous heliothine species H. virescens and H. armigera. This paper also provides an overview of the plant odour RN types identified so far in the three heliothine species (H. virescens, H. armigera and H. assulta). Altogether these results demonstrate that the olfactory RNs in heliothine species can be classified into distinct types, which correlate well with the principle of one receptor protein type expressed in each neuron of H. virescens females (Krieger et al. 2002). The RN types were functionally identified according to the compound eliciting the strongest response (the primary odorant) of which the most frequently recorded type of neurons in this study showed enantioselective responses to the acyclic monoterpene (+)-linalool. The primary odorant for the other RN types were (3Z)-hexenyl acetate, (+)-3-carene, E-pinocarveol, E-verbenol, vinylbenzaldehyde, 2-phenylethanol, methyl benzoate, a-caryophyllene and caryophyllene oxid.

Five of the RN types were found in the two species H. virescens and H. armigera. These types, like the five previously reported RN types (Paper I-III, Stranden et al. 2002, 2003), showed similarities that were noteworthy across the heliothine moths. Not only in the molecular receptive ranges and relative response strengths of primary and secondary compounds, but also the co-locations of RN types corresponded. This indicates that genes coding for important plant odorant receptors in the monophyletic heliothine species studied are conserved through evolution.

All compounds identified were known to be general constituents in several plant materials, e.g. floral compounds, oxidation products of the common monoterpenes a-and β-pinene, and aliphatic green leaf volatiles. Many of them are known as inducible, e.g. by caterpillar attack. Putative biological functions of the various odorants were discussed, either as attractants for nectar feeding or oviposition stimulants vs. repellents.

Discussion

Through the history of olfactory research, a central question has been which compounds activate the single RNs and can be considered as biologically relevant odorants in the various species. Even today, with the knowledge about the genes coding for olfactory receptor proteins, this question is still unresolved in most vertebrate and invertebrate species for food and plant odour information. When this study started, there was virtually no knowledge on how plant odour information is encoded in the RNs of heliothine moths. Many electrophysiological studies on insect olfactory RNs had been made by direct stimulation with synthetic compounds (review Masson and Mustaparta, 1990;

Dickens, 1990). Also in parallel with the present study results on RN tuning obtained by direct stimulation with selected odorants were made (Anderson et al., 1995; Jönsson and Anderson, 1999; De Bruyne et al., 1999; De Bruyne et al., 2001; Shields and Hildebrand, 2001, among others). Altogether, the various studies have reported broadly tuned RNs as well as RNs responding specifically to one or a few compounds.

However, tests with selected compounds are restrictive in it self, leaving the question open whether other compounds not tested might in fact be the biologically relevant odorants for the neuron. The method of GC-SCRs was used to test a single neuron for a large number of compounds, for instance sampled from the host plants by headspace collections.

The method of GC-SCR, first used in studies of the pheromone RNs (Wadhams, 1982; Löfstedt et al., 1982), were later employed for examining plant odour RNs (Tømmerås and Mustaparta, 1989; Wibe and Mustaparta, 1996; Blight et al., 1995; Stensmyr et al., 2001, 2003; Barata et al., 2002).

The improved GC-SCR method with two parallel columns, used in the studies of this thesis, allowed each neuron to be tested for the same mixture via two columns with different properties. This was an important upgrading since the active odorants often were found among the minor constituents, sometimes having overlapping retention times with other components.

Thus, the shift of retention times in the polar-and the non-polar columns was important for identifying the active constituents in the GC-MS analyses. After the successful use of the GC-SCR method, a further important step was made by exchanging one of the two columns with a column exhibiting chiral separation properties (Stranden et al., 2002). This was made to study the effect of pure enantiomers on the RNs, as shown in Paper III.

The results presented in the papers I-IV, as well as results of other studies in our laboratory, have demonstrated the importance of using this method of GC-SCR with two parallel columns for identifying biologically relevant plant odorants in insects (Stranden et al. 2002, 2003; Bichão et al., 2003; Bichão et al. in Press; Ulland et al., 2003).

Comparative Aspects

Comparative studies of the RNs in related species are important, since the results can provide information about conservation or changes of the functional properties through evolution. The results also become strengthened when the same RN types appear in more than one species. The plant odour receptor system in heliothines is particular interesting, since these species are considered as a monophyletic insect group, i.e. having a common origin (Matthews, 1999). The two species H. virescens and H. armigera, living on different continents, have been geographically separated for a long time and presumably exploited different host plant species, at least prior to the introduction of crop hosts they have in common. This suggests that changes between the species in sensitivity to plant odours might have evolved. However, the present studies have shown similarities of RN specificity that is noteworthy. All RN types found in more than one species showed remarkable similarities in molecular receptive ranges as well as in the relative sensitivity to the primary and secondary odorants as demonstrated by the dose-response relationships (Paper III, IV, Stranden et al. 2003).

The germacrene D RN type first classified in H. virescens (Paper I) and later in the other species (Stranden et al. 2002, 2003), showed striking similarities across the species, both concerning the response properties and as the most frequently occurring type. For all of them (-)-germacrene D had 10 times stronger stimulatory effect than the (+)-enantiomer, and the other compounds like (-)-a-ylangene elicited weaker responses (Stranden et al. 2002; 2003). Also the other RN types identified in two or three species were similar (Papers III-IV). Another interesting feature was the co-location of the same RN types in the three species, shown for the four types presented in paper II and III.

Similarities of the olfactory system within and between the heliothine species have also been found in the antennal lobe, by the invariance of number, size, form and position of the ordinary glomeruli (Berg et al., 2002; Skiri HT, Berg BG and Mustaparta H, submitted). Altogether, these studies suggest that both peripheral and some central features of the olfactory system are conserved in the three species of Heliothinae studied.

It is hypothesised that the species of the subfamily Heliothinae are evolved in a close relationship to agricultural host plants. Thus, the RNs might have been challenged by similar volatiles from the cultivated plants, which may have influenced RN specialisation during evolution. The question of which mechanisms make the heliothine species choose different host plants remains to be answered. Possibly this rely on species-specific olfactory RNs not yet identified or on differences in the central processing of odour information, if not, solely based on the contact chemoreception. In addition to the innate responses, the olfactory system has the capacity of plasticity as shown in experiments on olfactory learning and memory (e.g. reviews Menzel, 2001, Davis, 2004). It is hypothesised that previous experience might induce changes in the host preferences and thereby influence host-selection behaviour in heliothine moth, which could increase the utilization of abundant plants, like in monocultures (Firempong and Zalucki, 1991; Schoonhoven et al., 1998; Cunningham et al., 1999; West and Cunningham, 2002; Jallow et al., 2004). In studies combining appetitive olfactory learning and dual-choice wind tunnel tests, Cunningham et al. (2004) showed that H. armigera females trained on a certain odour preferred plants that were enhanced with the particular odour. However, olfactory learning does not seem to be the only mechanism influencing the different host plant choices. Laboratory experiments carried out with virgin females have shown that H. armigera and H. assulta choose different plants when given equal options (Wang et al., 2004).

To resolve the questions about the mechanisms underlying host plant selection, more studies are required. We know from studies of pheromone receptions in heliothine moths that RNs with similar specificity mediate different behavioural response in different species. According to this, the first step is to find out whether the identified plant odorants elicit similar or different behavioural responses in the females of the three species. So far, the primary odorant (-)-Germacrene D activating functionally similar RNs in all three species (Paper I, Stranden et al. 2002; 2003), has been shown to mediate attraction both of H virescens and H. armigera females (Mozuraitis et al., 2002; Gregg, personal communication). The aliphatic leaf odorant (3Z)-hexenyl acetate, which is the primary odorant of RN type

11 in H. virescens (Paper IV), is one of the compounds released by tobacco plants at night when infested with H. virescens larvae.

Thus, it might be involved in host repellence of H. virescens, preventing egglaying on plants already occupied by conspecific caterpillars (De Moraes et al. 2001). The presence of (3Z)-hexenyl acetate responding RNs in H. armigera is indicated by electroantennographic responses (Burguiere et al., 2001). In this species, (3Z)-hexenyl acetate mediate attraction to unmated H. armigera females when presented as a single compound in wind tunnel experiments (Gregg and Del Socorro, 2002). The opposite behavioural response observed in the two studies might either be due to species-specific differences of responses by different neurones to (3Z)-hexenyl acetate. Alternatively, mated and unmated females may respond differently. Possibly the repellence shown in H. virescens may in fact have been caused by other compounds. An interesting comparison of neuron specificity and behavioural responses across species has been made for the taste system of heliothine caterpillars.

The deterrent sensitive neurons of the taste sensilla (sensilla styloconica) of two heliothine caterpillars (Heliothis subflexa and H. virescens) showed no differences, neither in firing rate nor in adaptation to the taste stimuli tested (Bernays and Chapman, 2000). However, the behavioural threshold for rejection of toxic plant compounds (selected deterrents) during feeding was found to be lower in caterpillars of the monophagous H. subflexa than in those of H. virescens (Bernays et al., 2000). This implies a loss through evolution in the polyphagous H. virescens larvae to detect the compounds, after overcoming the toxicity. These results led to the conclusion that the different feeding behaviour of the two species were caused by different coding in the CNS rather than by differences in the peripheral sensory system. A similar principle may apply to the olfactory system.

When comparing RNs specificity in related species, one question is whether the same types are also present in unrelated species, which may point to a convergent evolution of the receptor proteins. RNs specialized for the same primary odorants as those found in heliothine species are also found in distantly related species. For instance RNs detecting E-β-ocimene has been shown in Spodoptera moths as well as in the two weevils Anthonomus grandis (cotton weevil) and Anthonomus rubi (strawberry weevil) (Dickens, 1990; Jösson and Anderson, 1999; Stensmyr et al., 2001; Bichão et al., in press). In some cases it is difficult to make a complete comparison of RN specificity because different test protocols have been used.

In the studies of the strawberry weevil A. rubi (Coleoptera: Curculionidae) and the lepidopteran moth Mamestra brassica similar test protocols as for the heliothine species were used (Bichão et al., in press; Ulland et al., 2003). Both similarities and differences of the RNs molecular receptive ranges appeared across these species. RNs tuned to linalool that were identified in all species (except H. assulta) responded secondarily to dihydrolinalool (Ulland et al. 2004). In contrast, other RNs tuned to the same primary odorants showed differences in sensitivity to the secondary odorants. The (-)-germacrene D RN type in the weevil (A. rubi) showed response to β-caryophyllene and no response to a-ylangene, whereas the opposite was the case for the (-)-germacrene D RNs of the heliothine moths (Stranden et al., 2002, 2003; Bichão et al., in press). The differences in molecular receptive ranges observed for some of the RN types in the distantly related species may reflect an independent evolution of the RN specificity during the adaptation to similar compounds of different host plants, or to chance mutation of common ancestral genes.

Coding of Odour Quality

The basis for recognition and discrimination of odour qualities in animals is the presence of RNs with different specificities for the odorants. In contrast to the visual system, operating with only three types of cones as bases for colour vision, the olfactory system is equipped with a much larger number of receptor protein types (Buck and Axel 1991). Each type is expressed in subsets of RNs, which projects to one or two glomeruli in the primary olfactory centre (reviewes Axel, 1995, Breer, 2003b). This implies a certain ratio between the number of glomeruli and the number of sensory neurons.

Whereas important processing of the visual information occurs in the retina, the olfactory information in vertebrates and insects is directly conducted by the RN axons to the primary olfactory centre of the brain. Certain principles of information processing typical in vision, like convergence on higher orders of neurons and lateral inhibition, is also important in the olfactory system, particularly studied in the primary olfactory centre (the AL of insects and the olfactory bulb of vertebrates). In insects, a large number of olfactory RNs converge on a smaller number of AL neurons (with a ratio of ~ 1000:1). Local AL inter-neurons provide lateral inhibition between glomeruli, which seems to be important in enhancing the contrast between active and inactive glomeruli (Smith and Shepherd, 1999).

Thus, activation of specific glomeruli by an odorant represents the code of the odour quality. This is particularly studied by optical recordings in honeybees, heliothine moths and the fruit fly (Galizia et al., 1999; Galizia et al., 2000; Galizia and Kimmerle, 2004; Skiri et al., 2004; Hallem and Carlson, 2004b). For interpreting results from this kind of studies, knowledge about the RN specificity is important and is in the present study provided for several RN types in heliothine moths.

The papers I-IV present 19 types of distinctly classified olfactory RNs in heliothine moths. However, a larger number is expected to be present, both indicated by the present electrophysiological data, and by the number of about 60 ordinary glomeruli in the antennal lobe of the three heliothine species studied (Berg et al., 2002; Skiri HT, Berg BG and Mustaparta H, submitted). With a ratio of 1:1 or 1:2 between the number of RNs and the ordinary glomeruli the heliothine antenna is expected to comprise in the range of 30-60 RN types. In future GC-SCR studies, we expect that additional RN types are recorded and classified according to their molecular receptive ranges. From the results of RNs so far obtained, the sharp tuning to one primary odorant and the low overlap of the molecular receptive ranges give the impression that the information about each odorant is mainly mediated by one RN type, similar to the pheromone system in these and other moth species.

This correlates well with the expression of one receptor protein type in each RN, as also indicated by the molecular biological study of olfactory gene expression in H. virescens (Krieger et al., 2002; 2004). Thus, the plant odour system in the heliothine moths seems to operate according to the principle of "labelledline" system, at least at low concentrations. However, this principle does not hold true for all the identified primary and secondary odorants. For instance (+)-linalool is the primary odorant for one RN type and a secondary odorant for another type (papers III and IV). A second example of overlap is between secondary odorants of two RN types responding to oxygenated bicyclic monoterpenes (Paper IV). Whether overlapping molecular receptive ranges is an important feature in the coding of plant odour information in heliothine moths remains to be seen in future studies, when the molecular receptive ranges of more RN types are identified. The relatedness of the few molecules out of hundreds tested, which activates the same type of RNs, support the principle that structurally similar molecules have a higher probability to bind to the same receptor proteins. This also explains the few cases of overlapping molecular receptive ranges within the same chemical group. This has also been shown in other studies of moths and weevils. Typical in all studies is that RNs of different chemical groups show no, or only minimal, overlap (Wibe and Mustaparta, 1996; Wibe et al., 1997; Stensmyr et al., 2001; Barata et al., 2002; Bichão et al., 2003). These results appear different from what is discussed according to results obtained in Drosophila, where a larger degree of overlap is found between the olfactory RNs (De Bruyne et al. 1999; 2001; Stensmyr et al. 2003; review Hallem and Carlson, 2004a).

The structure-activity relationships of the various RNs presented in this thesis indicate several molecular features of importance in receptor-ligand interaction. These are chirality, carbon chain length, electron dense parts and the flexibility of the molecules, which are reflected by enantiomers, number of C-atoms, double bounds and open vs. cyclic structures. These features are considered universal among receptorligand interactions in the olfactory system (Kafka, 1974; Priesner, 1977, 1979; Schneider et al., 1977; Bengtsson et al., 1990; Ohloff, 1986, 1994; Masson and Mustaparta, 1990, Mustaparta, 2002; Leal, 2001; Wibe et al., 1997, 1998; Borg-Karlson et al., 2003; Bichão et al., 2003; Bichão et al., in press; Laska, 2004, among others).

The results obtained in this thesis have been, and are currently, used in various other studies. This includes the use of optical recordings to study the representation of plant odorant qualities in the antennal lobe (Galizia et al., 2000; Skiri et al., 2004). Specific activity in distinct areas of the AL, mainly covering one or two glomeruli, has been shown for single odorants (Skiri et al., 2004). In addition, attempts have been made to trace the olfactory RN axons in the antenna lobe of heliothine females. These results show some correlation with the results from optical recordings (Paper III and Stranden et al. 2003). Thus, these preliminary results indicate that RNs responding to the same primary odorant project in one or a few glomeruli in the antennal lobe similar to what is found for the pheromone system in heliothine males (Berg et al., 1998; Berg, 1998). This is also in accordance with the current knowledge particularly from molecular studies in vertebrates and insects. For instance in Drosophila, the projections in one or two glomeruli are determined for RNs with identified genes coding for receptor proteins and described molecular receptor ranges (Keller and Vosshall, 2003; Hallem and Carlson, 2004a).

Ultimately, determining how odour information is coded in the brain requires linking a specific olfactory input with a behavioural output and correlating this with a measure of synaptic activity in the brain. This can be done in tests showing the ability of animals to discriminate between biologically relevant odorants, for instance by the use of the proboscis extension reflex in nectar feeding insects. In H. virescens, this reflex has been used to test some of the identified primary and secondary odorants (Skiri et al., in press), and these kind of studies are continuing for the other odorants identified in the present studies. As expected, Skiri and co-authors showed that H. virescens females, in a dose-dependent manner, were able to learn and to discriminate between linalool and both β-ocimene and β-myrcene, identified as primary and secondary odorants in paper III-IV. Surprisingly, the moths also showed the ability to discriminate between β-ocimene and β-myrcene, which in our experiments always activated the same RNs (Papers II, III).

These findings, which were supported by results from Ca2+-imaging experiments (Skiri et al., 2004) were explained by the possible presence of other RN types not yet identified, which responded to only one of the two odorants. In addition, impurities present in the samples at the relatively high concentrations tested might influence the discrimination. Also further processing of the olfactory information in AL projection neurons as well as higher orders of neurons (in the mushroom bodies and lateral protocerebrum) are important and may account for the discrimination of the two similar odorants (review Davis, 2004). For instance, in one study of the honeybee, synchronisation and temporal coding is suggested to be important in the discrimination of similar odorants (Stopfer et al., 1997).

Coding of Odour Intensity

The olfactory system seems to have the capacity to give information about odour intensity over several orders of magnitudes. The mechanisms involved in the coding of intensity can be ascribed to different response strengths of each RN to increased concentrations, to different sensitivity of each RN, as well as to central nervous mechanisms. The present studies present information about plant odour RNs types that is very sensitive to the primary odorants. In addition, the RNs within each type also show some variation in sensitivity. The best example comes from the frequently recorded (-)-germacrene D RNs, of which the most sensitive neurons responded to concentrations lower than the GC-detection limit, i.e. 1-10 pg (see Paper I) whereas higher concentrations are needed to activate other (-)-germacrene D RN. The doseresponse relationship showed that these neurons increase firing rates over 4-5 log units (Paper I, Stranden et al., 2003). Obviously, this intensity range can be transmitted directly to the AL neurons. Further increase of intensity might be provided by recruitment of RNs with lower sensitivity.

This mechanism has previously been suggested for detection of the major pheromone compound Z-11-16: AL in heliothine moths, having numerous sensilla along the antenna with RNs showing the same selectivity but different sensitivity to the same compounds. AL projection neurons of these moths respond with different sensitivity to antennal stimulation with the major component. This may be due to direct input from RNs with different sensitivity or to different numbers of RNs converging on each AL neuron. For the plant odour system the data that correlate sensitivity of the RNs to the sensitivity of the projection neurons are scarce. In general, a low sensitivity is observed for the projection neurons responding to antennal stimulation with plant odorants (Roche King et al., 2000; Anton and Hansson, 1995; Greiner et al., 2002; Masante-Roca et al., 2002; Reisenman et al., 2004).

This may be due to down regulation of the neurone sensitivity, by modulation of serotonin or octopamin (Kent et al., 1987; Sun et al., 1993; Kloppenburg and Hildebrand, 1995; Mercer et al., 1996). Alternatively, the responses recorded from the projections neurons are not ascribed to stimulation with the primary odorant of the RNs giving the input. Results from optical recordings experiments (Ca2+-imaging) in various species, including heliothine, shows that an increasing number of glomeruli were recruited with increasing odour concentrations (Sachse and Galizia, 2003; Carlsson and Hansson, 2003; Skiri et al., 2004). In the honeybee, this increase seems to be due to overlapping molecular receptive ranges of the RNs. A general assumption is that both RN sensitivity and the total number of RNs tuned to a particular odorant are important for the distance over which the odorant is detected. The large number of very sensitive RNs responding to germacrene D in H. virescens indicates that this compound may play a significant role over a long distance, probably in attraction (Mozuraitis et al., 2002). Recruitment of (-)-germacrene D RNs with lower sensitivity may be activated closer to the odour source, and give additional information about the intensity at short range.

Behavioural Implications

During recent years, attention has been given to induction of compounds emitted by a plant. Both biotic and abiotic factors influence the production and release of volatiles. Interesting biotic factors are volicitin and structural analogue compounds present in caterpillar salvia (Alborn et al., 1997; Mori et al., 2001, 2003). During feeding, these are suggested to induce production and release of certain compounds like (3Z)-hexenyl acetate, (3Z)-hexenol, E-β-ocimene, linalool, E-4,8-dimethyl-1,3,7-nonatriene, E-β-farnesene, and E,E-a-farnesene (Loughrin et al., 1994; McCall et al., 1994; Röse et al., 1996; De Moraes et al., 2001; Heiden et al., 2003; Röse and Tumlinson, 2004). It is also assumed that the induced release of volatiles is partly due to the constant cutting of leaf tissue during feeding. However, both qualitative and quantitative differences of the induced volatile profiles appear by feeding of different moth species, including the heliothines (Mori et al., 2001; Röse and Tumlinson, 2004; De Moraes and Mescher, 2004). This indicates that species specific factors in the salvia influence the induction. Compounds induced by H. virescens caterpillars have been suggested to mediate host repellence, preventing mated conspecific females to lay eggs on infested plants (De Moraes et al., 2001). This repellence behaviour might be mediated by several of the RN types identified in this study (Papers II-IV); detecting E-β-ocimene [secondary odorant E-4,8-dimethyl-1,3,7-nonatriene], (3Z)-hexenyl acetate and (3Z)-hexenol, E,E-a-farnesene [secondary odorant E-β-farnesene], (+)-and (-)-linalool, respectively. The enantiomeric ratio of linalool emitted during caterpillar feeding is not reported.

To become an insect repellent signal, the compound produced by the direct defence mechanisms in infested plants might be associated with increased concentrations of toxins or a lower nutrition value, having a negative effect on the development and survival of the offspring (Mori et al., 2001). The fitness advantages to herbivores avoiding oviposition on induced plants are obvious, as these plants are likely to host not only larvae that represent potential competitors for the offsprings, but may also attract natural enemies. Parasitoids and predators of the heliothines have the ability to learn to discriminate between plant emitted volatiles that are induced by different species of caterpillars (Meiners et al., 2002, 2003). Two of the compounds commonly induced by heliothine caterpillars (e.g. H. virescens), linalool and β-ocimene, are attractive to various parasites and predators (De Moraes et al., 1998; De Moraes and Mescher, 2004, Röse and Tumlinson 2004). The two compounds are not induced in the host plant Physalis angulata when attacked by caterpillars of the monophagous Helicoverpa subflexa (De Moraes and Mescher, 2004). This is explained by the adaptation of H. subflexa to monophagy on Physalis fruits, lacking linolenic acid. Other plant species contain this acid, which is required both for development and morphogenesis of most insects, and important for the induction of linalool and β-ocimene in plants (De Moraes and Mescher, 2004). In this way, H. subflexa caterpillars feeding of Physalis fruits exhibit a clear competitive advantage as compared with H. virescens, by overcoming the lack of linolenic acid and not inducing attraction to the predators (De Moraes and Mescher 2004).

Plants emit blends of volatiles that vary both qualitatively and quantitatively in different parts of the plant, as well as with age or during the diurnal or seasonal cycles. Herbivore insects might exploit signals specific for these conditions when searching for a host. In nocturnal emission of flowering tobacco plants, it is found a four-fold increase in the amount of aromates like 2-phenylethanol, methyl benzoate and benzaldehyde (Raguso et al., 2003). It was suggested that these floral compounds are produced for attracting night active pollinators and may also serve as cues for the noctuid herbivorous moths in their search for nectar (Raguso et al., 2003).

This is for instance shown for 2-phenylethanol tested in a two-choice olfactometer (Gregg and Del Socorro, 2002). Many other compounds identified as primary and secondary odorants for the heliothine species are shown to be attractive to mated or unmated females in various behavioural bioassays. These include linalool, 3-carene, geraniol, a-caryophyllene, β-caryophyllene and (-)-germacrene D, which were either tested as single compounds or constituents added to blends (Rembold and Tober, 1985; Rembold et al., 1991; Jallow et al., 1999; Bruce and Cork, 2001; Hartlieb and Rembold, 1996; De Moraes et al., 2001; Mozuraitis et al., 2002; Gregg and Del Socorro, 2002; Robert Heath, personal communication). (-)-Germacrene D is particularly interesting because of the numerous RNs found on the antenna of the three heliothine species. This compound, tested in two independent studies, has been indicated to act as an oviposition stimulant and/or attractant for mated H. virescens, and as an attractant for virgin H. armigera females (Mozuraitis et al., 2002; Peter Gregg, personal communication).

Concluding Remarks

Striking similarities were found when comparing the RN specificity of the three heliothine species. Furthermore, the RNs were tuned to general plant constituents of which many seems to be related to certain condition of the plant e.g. caterpillar attack. Because these species are able to utilize a broad range of host plants, it is possible that other RN-types might be revealed when including more plant species. More electrophysiological data, in particular on the less studied oligophagous H. assulta, is required for detailed comparisons of the RN types. In addition, other species of the subfamily heliothine may be included in future studies. H. subflexa is particularly interesting because of its specialisation on Physalis fruits. Although our results are complementary to other studies showing behavioural significance of some of the odorants identified, more behavioural studies are required to elucidate the biological role of the various odorants. Interesting objectives are studies concerning the ability of the moths to learn and to discriminate single components and mixtures of the identified odorants.

The results obtained in the present studies have shown 19 types of olfactory RNs of which primary and several secondary odorants are identified for 16 of them (Paper IIV, Stranden et al. 2002, 2003). These data may be used in future studies of the peripheral olfactory events, including odorant-receptor interactions and transduction mechanisms. This requires molecular biological characterisations of olfactory genes and receptor proteins in heliothine moths, studies that are in progress by Krieger et al. (2002; 2004). The identified odorants are also used in studies of the central processing of olfactory information and olfactory learning, carried out in our laboratory. In addition, the use of plant odorants in integrated control of heliothine moths makes the present results interesting also for applied research.

3

Olfactory Receptor Neurones Specified

INTRODUCTION

The chemical senses are considered as the phylogenetically oldest of the senses. Unicellular organisms can orient along gradients of toxic and nutritional chemicals. More complex organisms have evolved senses based on advanced neural systems for detection and processing of information about the external chemical stimuli; olfaction for airborne molecules and taste for molecules dissolved in liquid. In addition, a sense for noxious molecules in air and liquid is present. Separate sensory organs contain the receptors that detect the stimuli and mediate the information further to different brain areas via interneurones. In spite of the interest and fascination about chemical stimuli throughout the history of human cultures, little knowledge, particularly about the olfactory mechanisms, has existed until recent years.

Whereas visual and auditory stimuli are well defined by one feature (the wavelength), a chemical stimulus has several interactive features that make it difficult to screen for a continuum of important molecular characteristics. Thus, the problem is to find the biologically relevant odorants among the hundreds of known and unknown compounds surrounding the organisms. The identification of relevant odorants for olfactory receptor neurones (RNs) has not become less important in the recent years of olfactory research. The impressive development in molecular biological studies has resulted in the identification of numerous genes in various species coding for olfactory receptor proteins. At present, one key question is which odorants the receptor proteins are specified for. This question can only be answered by physiological studies.

The Olfactory System

The anatomy of the olfactory system is well known in many species. Although the location of the olfactory epithelia in vertebrates (upper nasal cavity) is very different from the location in insects (in sensilla on the antennae), the principal anatomy of the two systems is strikingly similar (reviewed by Steinbrecht 1992, Hildebrand and Shepherd 1997). The ciliar parts of the primary sensory neurones, containing the receptor proteins, are embedded in a protein-rich lymph. The axons of the RNs project directly to the primary olfactory centre of the brain, the olfactory lobe in vertebrates and the antennal lobe in insects. In these centres, the primary axon terminals synapse with the second order of neurones in the typical spherical glomeruli (Hanström 1928, Boeckh et al. 1990, Leise 1990, Boeckh and Tolbert 1993) that seem to be conserved in number and location between individuals of the same species (Allison 1952, Meisami 1991, Rospars 1988). There are two separate olfactory systems, one for species-specific chemical signals, e.g. pheromones, and one for other odorants, e.g. food odours. The easily accessible olfactory system in insects and its similarity with the vertebrate olfactory system makes insects suitable model organisms for studying the mechanisms of olfaction.

Olfactory Sensilla, Perireceptor and Transduction Events

The olfactory organs, olfactory sensilla, in insects are usually located on the antenna. Many lepidopteran species have sexually dimorphic sensilla. For instance, the sensilla trichodea type I in heliothine moths, containing the pheromone RNs, are only present in male antennae, whereas s. trichodea type II and s. basiconica are present in both sexes and mediate plant odour information (Almaas and Mustaparta 1990, Almaas and Mustaparta 1991, Koh et al. 1995). In addition, s. trichodea type II also mediates pheromone information in males (Almaas and Mustaparta 1991). The different sensilla types are classified according to internal and external anatomy (Schneider 1964, Ernst 1969, Schneider and Steinbrecht 1968, Keil 1984, Steinbrecht 1992, Keil et al. 2001).

The outer structures are often hair formed, with cuticle walls perforated by pores allowing hydrophobic volatiles to enter the sensilla (Ernst 1972). The internal cavity of the sensilla is filled with the receptor lymph, which constitutes a hydrophilic barrier for the odorants. The lymph surrounds the dendrites of the olfactory neurones, which possess the membrane spanned receptor proteins. Odorant-binding proteins (OBPs), present in the receptor lymph, show selective binding to some odorants and are suggested to function both as a filter and as a carrier, transporting the odorants through the lymph to the receptor proteins (Prestwich et al. 1995, Steinbrecht 1998).

These suggested functions are supported and further enlightened by recent experiments which show that pheromone-binding proteins (PBPs) in moths bind to several molecules, but conformational changes only occur in interaction with the specific pheromone (Campanacci et al. 2001, Bette et al. 2002, Mohl et al. 2002). The OBPs might also contribute to the inactivation of the odorant, as discussed by Kaissling (reviewed in 1998). The identification of OBPs in insects started out from antennal homogenates of the giant silkmoth Antheraea polyphemus (Vogt and Riddiford 1981). Afterwards OBPs, mainly PBPs, have been identified in several insect species (reviewed by Vogt et al. 1999). The number of OBP types identified in each species has remained small until recent studies of the fruitfly Drosophila melanogaster genome revealed 39 genes coding for putative OBPs (Shanbhag et al. 2001). This indicates the presence of several non-pheromonal OBPs binding host odorants.

The olfactory neurones are activated by the binding of the odorants or the complex odorant-OBPs to the membrane receptors. By activation, an intracellular G-protein coupled cascade reaction is induced, which produces the second messenger inositol 1,4,5-trisphosphate (IP^3) or adenosine 3,5-monophosphate (cAMP) (reviewed by Breer et al. 1990b, Paysan and Breer 2001, Krieger and Breer 1999). Evidence for the IP^3 pathway in insect olfactory RNs is based on biochemical and electrophysiological experiments. The rapid increase of the concentration of IP^3 upon stimulation with pheromones occurs within the time of the receptor potential (Breer et al. 1990a). Furthermore, the transduction depends on the presence of G-proteins, which are identified in insect antennae (Breer et al.

1990a, Laue et al. 1997), and on the opening of cation channels by IP^3 , which is demonstrated in insect olfactory neurones (Stengl 1994, Wegener et al. 1997). The cAMP pathway may also be important in insects (Krieger and Breer 1999). This is indicated by the expression of cAMP sensitive channels in the antennae (Baumann et al. 1994, Krieger et al. 1999). It is suggested that this pathway is required for information about a subset of odorants (Dubin et al. 1998).

Expression of Receptor Proteins in Single Neurones and the Principle of Convergence in the Antennal Lobe

The first identification of a multigene family that might encode odorant receptor proteins was made in the rat by Buck and Axel (1991). This extremely large and divergent gene family was expressed exclusively in the olfactory tissue. The genes were further characterised by a central domain encoding seven transmembrane regions, known to be a typical feature of G-protein coupled receptors.

Since then, putative genes for receptor proteins have been identified in many species of vertebrates and invertebrates, e.g. 1000 types in the mouse and 61 in the fruitfly (reviewed by Mombaerts 1999, Keller and Vosshall 2003). Functional evidence is obtained for four olfactory receptor genes, one in the human (Wetzel et al. 1999), one in the rat (Zhao et al. 1998), one in the mouse (Touhara et al. 1999) and one in the fruitfly (Störtkuhl and Kettler 2001, Wetzel et al. 2001).

The total number of gene types tested for expression is small, but so far, little co-expression of different receptor types has appeared in the neurones. Two exceptions are found; one type present in all olfactory RNs of the fruitfly is believed to possess a non-olfactory function (Keller and Vosshall 2003), and co-expression of the two olfactory receptor mRNA is found in a subset of olfactory RNs in rats (Rawson et al. 2000).

Thus, the current held belief is that a single olfactory receptor gene type is expressed in each olfactory neurone. An exception is in the soil nematode Caenorhabditis elegans, which on average express 20 gene types coding for receptor proteins in each of the 32 chemosensory neurones (Troemel et al. 1995, 1997, reviewed by Bargmann 1998).

Olfactory RNs expressing one type of receptor proteins are in most insect species distributed over the whole antennae (i.a. Almaas and Mustaparta 1990), and in vertebrates over one of the four zones of the olfactory epithelium (Ma and Shepherd 2000, reviewed by Mombaerts 1999). Still the widely distributed neurones of the same type converge onto one or a few glomeruli, as shown both in vertebrates and invertebrates (Ressler et al. 1994, Vassar et al. 1994, Mombaerts et al. 1996, Gao et al. 2000, Vosshall et al. 2000). The ratio of convergence in insects is about 100-1000:1 and in vertebrates 25.000:1 (in rabbits) (Rospars 1988, Allison 1952). Staining of active insect pheromone RNs of specific types, supports the convergence principle (Berg et al. 1998, Berg 1998). However, exceptions from this rule have also been reported (Hansson et al. 1988, Anton and Hansson 1999). The organised convergence in insects starts before the axons grow into the antennal lobe, as an interaction between the single fibres and the glia cells, which is shown at the late metamorphic stage in the development of the antennal lobe (Rössler et al. 1999).

Encoding of Odour Information in Receptor Neurones

A rough overview of odorants eliciting electrical responses across the olfactory epithelia can be obtained by recordings of electro-antennogram (EAG) in insects and electro-olfactogram in vertebrates (EOG) (Schneider 1955, Ottoson 1956, 1971). To determine which of the odorants activate a single olfactory RN (the molecular receptive range of the neurone), responses to the odorants are recorded from single cells by placing a glass capillary or tungsten microelectrode into the base of one sensillum hair (Schneider 1957a,b, Boeckh 1962), or a glass capillary microelectrode over a cut hair ("tip-recording") (Kaissling 1974). Trying to characterise the RNs in respect to specificity, one faces two major problems. The naturally produced odorants exist in blends of volatiles, and commercially available compounds are rarely 100% pure. It is important to use highly pure compounds for stimulation; otherwise the response of the RNs might be caused by the impurities in the samples tested. Obviously, testing naturally produced volatiles by direct stimulation does not show which of the constituents are active. A fruitful method is the linkage of high resolution gas chromatography to single cell recordings (GC-SCR), first carried out in pheromone analysis by Wadhams (1982), and later for analysing plant odorants (Tømmerås and Mustaparta 1989, Blight et al. 1995, Wibe and Mustaparta 1996). With this method, constituents of a mixture of volatiles can be presented to the RNs separately.

Early studies of olfactory RNs, carried out by testing commercially available compounds, has resulted in different principles on how odour information is encoded. Based on these studies, Vareschi (1971) defined two gross categories of RNs. One category was called "odour specialists", where neurones are highly specialised for one compound and show the same specificity, i.e. the same dose-response relationships. The other category was termed "odour generalists", i.e. RNs responding to many compounds and showing significantly different and overlapping "reaction spectra" (molecular receptive range). The pheromone RNs fulfil the requirement of "odour specialists" (i.a. reviewed by Schneider 1969, 1992, Hansson 1995, Mustaparta 1996), also according to another definition that demands a "biological importance (relevance)" of the odorant (Schneider and Steinbrecht 1968, Kaissling 1974).

In the early studies, the plant odour RNs were categorised as "odour generalists", but showed different sub-divisions in various species, including some narrowly tuned neurones (reviewed by Masson and Mustaparta 1990). The question about biological relevance of the plant odorants came more into focus by testing naturally produced odours and trying to isolate the active compounds.

The possibility that key compounds exist for different food types was ruled out in studies of the cockroach Periplaneta americana (Sass 1978, Selzer 1981). Here, stimulation with fractions of lemon oil revealed seven RN types responding to different fractions. The major component limonene, did not elicit responses in any of the RNs. Recent evidence for plant odour RNs being as sensitive and selective as pheromone RNs as well as RNs responding to several compound classes, is taken from studies using direct stimulation (Dickens 1990, Hansson et al. 1999, Jönsson and Anderson 1999, Larsson et al. 2001, Shields and Hildebrand 2001) and particularly from studies employing GC-SCR (Wibe and Mustaparta 1996, Røstelien et al. 2000a,b, Stensmyr et al. 2001, 2003, Barata et al. 2002, Bichão et al. 2003). The latter studies exclude responses to the numerous inactive plant constituents and frequently show strongest responses to minor components. This stresses the importance of using GC-SCR for identifying plant odorants.

Based on the specificity of the studied RNs, two principles have emerged on how the odour information is mediated from the periphery to the central nervous system (CNS). One principle called "labelled line", is used in chemical senses to express that information about one compound is mediated to the brain by one type of RNs, i.e. it involves no overlap of molecular receptive ranges between different RN types (Shepherd 1984). The other extreme principle called "across-fibre pattern mechanism", means that information about one compound is mediated by different RN types, i.e. with overlap of molecular receptive ranges and different sensitivity to the compound (Gesteland et al. 1965). The mediation of pheromone information from the periphery to the antennal lobe in insects seems to follow the principle of "labelled line".

Furthermore, the RNs belonging to each type project into one of the male specific glomeruli (Berg et al. 1998, Berg 1998, Galizia et al. 2000b), constituting the macro-glomerular complex (MGC) that receives information about insect produced signals (Boeckh and Boeckh 1979, reviewed by Boeckh and Tolbert 1993). Information about food and host odours is commonly believed to be mediated by the more complex mechanism of "across-fibre pattern" (reviewed by Lemon and Getz 1999), in spite of the fact that narrowly tuned RNs are found. Projections by functional types of RNs in the numerous "ordinary" glomeruli remain to be determined. However, optical recordings have shown glomerular maps of odour qualities that are conserved between individuals (reviewed by Galizia and Menzel 2001).

Olfactory Coding and Olfactory Learning

The antennal lobe has two major groups of neurones, the local inter-neurones and the projection neurones forming a network with the synaptic terminals of the RNs (reviewed by Chapman 1998, Anton and Homberg 1999). The GABAergic local inter-neurones innervating many glomeruli, are thought to mediate important inhibition in the antennal lobe, e.g. to increase the contrasts between activated and non-activated glomeruli (Homberg et al. 1989, Joerges et al. 1997, Galizia et al. 2000a, Sachse and Galizia 2002) and inhibit activation between odour pulses of pheromones (Christensen et al. 1993). The projection neurones conduct the information to higher brain centres, the mushroom bodies and the lateral protocerebrum. The mushroom bodies are shown to play a significant role in learning and memory of odours (reviewed by Menzel 1999, De Belle and Kanzaki 1999, Heisenberg 2003), while the lateral protocerebrum is a premotoric area, from which descending neurones make synapses with motor neurones in the suboesophageal ganglion and the thoracic ganglia (Strausfeld 1976).

A challenging question in odour information processing concerns how olfactory information is coded spatially and temporally, leading to odour discrimination. Since each glomerulus seems to receive information from one type of RNs, the next question is whether there is a correspondence between the input and the output of the glomeruli in the antennal lobe. This is found for the MGC of H. virescens, by intracellular recordings of antennal lobe neurones (Christensen et al. 1995, Vickers et al. 1998, Berg et al. 1998, Berg 1998). It means that projection neurones responding to antennal stimulation with an odorant have at least one dendrite branch in the glomerular unit receiving the information.

However, in another species a mismatch between input and output information is shown in the antennal lobe (Hansson et al. 1994, Anton and Hansson 1999). More data on different species are needed in order to draw general conclusions about this principle. Intra-and extracellular recordings and multi-unit recordings of projection neurones have shown that different blends of plant volatiles activate different sub-populations of projection neurones, indicating a spatial representation of odour qualities (Laurent and Davidowitz 1994, Laurent et al. 1996, Christensen et al. 2000). In the locust and the honeybee (Apis mellifera), each projection neurone shows an odorant specific slow temporal pattern (Laurent et al. 1996).

In addition, a fast temporal pattern is recorded as local field potential oscillation cycles, ascribed to activation of different projection neurone populations. The oscillations, caused by the GABAergic local interneurones, are shown to be important for discrimination of structural similar odorants in the honeybee (Stopfer et al. 1997). It is speculated whether the observed oscillatory synchronisations may provide coding strategies for odour concentration and odour quality (reviewed by Stopfer et al. 1999, Laurent 1999, 2002, Galizia and Menzel 2000). Another study of projection neurones in a moth species, using stimulation with brief pheromone pulses, revealed a temporal pattern of synchronous activity that was dependent on odour intensity and intermittency (Christensen et al. 2000). However, the temporal pattern was not found to be oscillatory or odour specific. Whether these differences can be ascribed to different coding principles in different insect species or in different olfactory systems (general odorants versus pheromones) remains to be clarified.

An important aspect of olfaction is learning of odours. In order to recognise an odour, both neuronal detection and memory of the odour are necessary. An interesting question is-What are the neural mechanisms underlying learning and memory of odours? The importance of the mushroom bodies in olfactory learning and memory is well described in the honeybee by classical conditioning experiments using appetitive learning (reviewed by Menzel 1999, Müller 2002). The conditioned stimulus (odour) is associated with the unconditioned stimulus (sucrose stimuli and reward) that elicits the proboscis extension reflex (PER).

One neurone, an octopaminergic ventral unpaired median neurone (VUMmx1), is shown to mediate information between the olfactory and the taste pathway in the learning process (Hammer 1993). Another neurone projecting from the mushroom bodies to the lateral protocerebrum, the pedunculus extrinsic neurone (PE1), is shown to change the responses after learning experiments (Mauelshagen 1993). The cellular and molecular mechanisms are enlightened by biochemical studies in the honeybee and by molecular biological studies of the fruitfly (reviewed by Menzel 1999, Heisenberg 2003). Appetitive learning of olfactory information is also shown in lepidopteran species, including heliothine moths (Hartlieb 1996, Fan et al. 1997, Daly and Smith 2000, Daly et al. 2001a,b).

The Nature of Plant Odorants

The large number of diverse compounds produced by plant species is well known from research in natural product chemistry as well as in chemical ecology. Plant compounds are built from products of the photosynthesis in diverse biosynthetic routes catalyzed by various enzymes. The compounds essential for growth and development of the plant are called primary metabolites, and others secondary metabolites (Hartmann 1996). The same compound may have both a primary and a secondary role. Secondary metabolites were first described as by-products of the plant metabolism by Sachs in 1873, and were traditionally considered as metabolic wastes and detoxification products (Haslam 1986, Luckner 1990).

Today these compounds are classified as an essential part of the plant biochemical equipment to cope with the environment (Fraenkel 1959, Hartmann 1996, Pickett et al. 1999). Ehrlich and Raven (1964) were among the first to suggest that secondary metabolites evolved in a co-evolutionary arms race of plant defences against herbivory and herbivore animal responses. The plants produce and accumulate toxins, which is followed by a reciprocal response in the insect to detoxify or excrete the toxins. Being toxic also to the plant themselves, the secondary metabolites are often produced as pro-toxins (i.e. enzymatically ignited upon attack), constitutively accumulated in special organs (e.g. vacuoles, glandular trichomes) or induced (i.e. produced in response to attack) (Hartmann 1985).

The blend of volatiles produced by a plant is a complex mixture, often consisting of hundreds of compounds. The blends can be collected by various methods of distillation, headspace collection and extraction (reviewed by Silverstein and Rodin 1966, Agelopoulos and Pickett 1998, Millar and Sir 1998). The trapped volatiles are analysed by separation in a gas chromatograph, and the compounds are identified by mass spectrometry, followed by confirmation with authentic materials. Continually more plant constituents present in nature are identified, due to more sensitive analytical methods employed. Plant volatiles belong to many different chemical groups like short chain alcohols, aldehydes, ketones, esters, aromatic phenoles and lactones as well as mono-and sesquiterpenes (reviewed by Gibbs 1974, Smith 1976, Bernays and Chapman

1994, Ohloff 1994, Schoonhoven et al. 1998). Some compounds are mainly found in restricted plant taxa, like the isothiocyanates in Brassicacea (reviewed by Kjær 1976, Fahey et al. 2001). Others are commonly occurring, like the "green leaf volatiles" (mainly six-carbon alcohols and aldehydes) that are products of the lipid metabolism in green leaves (Visser and Avé 1978, Hatanaka 1993). Flower fragrances, constituting hundreds of compounds, may include different molecules than those of the vegetative plant parts (reviewed by Dudareva and Pichersky 2000). The quantity and quality of the volatile blends released by plants are affected by both biotic and abiotic factors (i.a. Paré and Tumlinson 1999, Gouinguené and Turlings 2002).

Most of the plant constituents are produced via a few basic biogenetic routes, leading to one or a few key-metabolites, from which numerous derivatives are formed, usually by simple enzymatic transformations (Hartmann 1996). For instance, geranyl diphosphate is the unique precursor of all monoterpenes and farnesyl diphosphate of all sesquiterpenes (reviewed by Dewick 1999, Rohmer 1999). The production of compounds is often temporally and spatially restricted in the plants by the expression of genes coding for the enzymes, as shown for de novo biosynthesis of floral volatiles (Dudareva and Pichersky 2000). Furthermore, the enzymes can be enantiospecific, producing only one of the two enantiomers of a molecule (Schmidt et al. 1998, 1999, reviewed by Ohloff 1994).

Heliothine Moths

Plant produced volatiles are important cues for the herbivorous insects, including heliothine moths, in the location of their hosts for feeding and oviposition (Rembold and Tober 1985, Tingle et al. 1990, Mitchell et al. 1991, Tingle and Mitchell 1992). The emitted plant volatiles are transported by turbulent air that breaks the signal up into discrete packages with increasing distance from the source (Murlis 1986). Females searching for host-plants fly upwind by detection of attractants present in the air plume, i.e. showing positive anemotaxis toward the odour source (Tingle et al. 1990, Mitchell et al. 1991, Tingle and Mitchell 1992). After making contact with a plant, acceptance or rejection of the plant is made also on the basis of information mediated by other senses; taste and mechano-sensation (reviewed by Bernays and Chapman 1994, Schoonhoven et al. 1998). Experience of a host plant visit seems to affect later choices of host plants in female heliothine moths (Firempong and Zalucki 1991). It was shown that experience with a host plant significantly increased the probability of selecting that plant species for subsequent oviposition and feeding (Cunningham et al. 1998a,b). Thus, it is speculated whether learning is involved in the mechanisms of host plant selection (Cunningham et al. 1999).

The subfamily Heliothinae (Insecta; Lepidoptera; Noctuidae) consists of more than 80 different species, of which several are considered as world-wide major pests on crops, like the American tobacco budworm moth Heliothis virescens (Fabricius) and the cotton bollworm moth Helicoverpa armigera (Hübner) (Todd 1978). H. virescens, is distributed over the American continent and uses as principal hosts, i.a. plants of tobacco, cotton, tomato, sunflower and soybean (Fitt 1989, Matthews 1991). H. armigera, geographically separated from H. virescens, is distributed over South Europe, Asia, Australia, Africa and Eastern Pacific. As a principal host, this species uses plants of i.a. maize, sorghum, sunflower, cotton, tobacco, soybean, pulses (chick pea, pigeon pea), safflower, rapeseed and groundnuts (Fitt 1989, Matthews 1991). H. armigera has been found on at least 60 cultivated and 67 wild host plants in 39 families (reviewed by Reed and Pawar 1982, Zalucki et al. 1986). The major pest status of the two heliothine species are explained by their high polyphagy, mobility, fecundity and the ability to have facultative diapauses, in addition to the ability to resist different pesticides (Fitt 1989). The Oriental tobacco budworm Helicoverpa assulta (Guenée), partly sympatric with H. armigera, is distributed over Asia. This species, considered oligophagous, has a relatively narrow host plant range, mainly feeding on plant species of Solanacea such as tobacco, tomato, hot pepper and several Physalis species (Hill 1983, Matthews 1991, Park 1991).

Olfaction in heliothine moths has been studied in several species, particularly as it concerns pheromone communication. Chemical analyses combined with behavioural studies have revealed which compounds and ratios of them act in intra-as well as interspecific communication (i.a. Kehat et al. 1980, Vetter and Baker 1983, Kehat and Dunkelblum 1990, Cork et al. 1992, Boo et al. 1995, Vickers and Baker 1997). The detection of these chemical signals by the RN as well as the processing of the information in the antennal lobe has been studied (reviewed by Mustaparta 2002). However, the knowledge of encoding and processing of other odorants have been scarce in heliothine moths as well as in other organisms. Until recent GC-SCR studies by Røstelien et al. (2000a,b), little knowledge has existed about plant odorants in heliothine moths. The objectives of the present thesis were to identify important plant odorants detected by females of the three species, H. virescens, H. armigera and H. assulta, and to determine how the information is encoded in the RNs.

AIMS OF THE THESIS

The aims of the thesis were:

1. To identify naturally produced plant volatiles that are detected by receptor neurones in females of the heliothine moths Heliothis virescens, Helicoverpa armigera and Helicoverpa assulta.
2. To determine the molecular receptive range and sensitivity (specificity) of the plant odour receptor neurones.
3. To compare the specificity for functional classification of the plant odour receptor neurones within and between the three related heliothine species.
4. To explore the possibility of staining plant odour receptor neurones for determining the projection pattern in glomeruli in the antennal lobe.
5. To assess the biological importance of one odorant activating a major type of plant odour receptor neurones in females of one species (Heliothis virescens).

SURVEY OF THE INDIVIDUAL PAPERS

In the papers forming the basis of this thesis, efforts have been made to elucidate the five aims given above. Five RN types have been studied in three related heliothine species, concerning specificity. Preliminary data on the projection of plant odour receptor neurones is obtained. For one important odorant, the behaviourally significance is revealed.

Paper I

Paper I was based on the findings by Røstelien et al. (2000a) that H. virescens females have one major type of plant odour RNs responding to the sesquiterpene germacrene D. By electrophysiological recordings linked to gas chromatography, single RNs in females of the related species H. armigera were found to respond with high sensitivity and selectivity to germacrene D. Germacrene D was present as a constituent in headspace or in extracts of different plant species, in fractions isolated from essential oil by medium pressure liquid chromatography (MPLC) or in reference samples (provided by Prof. W. A. König).

Since germacrene D consists of two enantiomers, the aim was to determine whether these neurones responded selectively to one enantiomer or to both. Optical enantiomers in general and those of germacrene D in particular are difficult to obtain as 100% optically pure compounds.

Therefore, this study was performed by the use of a chiral GC-column (provided by Prof. W. A. König), which could separate the germacrene D enantiomers. By installing this column in parallel with a non-chiral column in the GC, the RNs could be tested for mixtures and fraction of plant volatiles via both columns.

When germacrene D was injected into the chiral column, the two enantiomers were well separated; (+)-germacrene D eluted before (-)-germacrene D. The results obtained indicated that all germacrene D RNs belonged to the samc type; i.e. all responded to both enantiomers, but (-)-germacrene D had about 10 times stronger effect than the (+)-enantiomer. Parallel dose-response curves for the two enantiomers were obtained by direct stimulations.

The enantiomeric composition of germacrene D, determined on the basis of the neurone responses, was found to differ in six plant species and in different individuals of one species. The results, showing the presence of only one neurone type for receiving the information about germacrene D in the various plants, suggest that the two enantiomers mediate the same kind of information to the moth brain, but with different intensity.

Paper II

The aim of the study in paper II was to find out whether RNs responding to germacrene D might be present in more of the heliothine species, and whether RNs responding to the same compounds exploit the same specificity, i.e. whether they show similar structure-activity relationships. Females of the three related species H. virescens, H. armigera and H. assulta were studied. The presence of germacrene D RNs was shown in H. virescens by Røstelien et al. (2000a) and in H. armigera in paper I. In paper II, germacrene D RNs were also identified in H. assulta. In addition, structure-activity data of the germacrene D RN type was revealed in the three species. Altogether 69 recorded RNs were classified, based on 436 GC-SCR using five different column types, and included up to 37 recordings from the single neurones.

The RNs in the three species showed similar ranking of responses to the nine active sesquiterpenes. (-)-Germacrene D was about 10 times more effective than the (+)-enantiomer, being again 10 times more effective than (-)-a-ylangene. Weaker effects were obtained for (+)-a-copaene, one unidentified sesquiterpene (no. 3), (+)-β-ylangene, β-copaene and both enantiomers of one unidentified sesquiterpene (no. 6). All active compounds were structurally similar to (-)-germacrene D. Comparing the structures of the active molecules, shows that the important properties of (-)-germacrene D are the flexibility of the ten-membered ring system and the three double bonds acting as electron rich centres, as well as the direction of the isopropyl group. The narrowly tuning of the RNs was demonstrated by the lack of activity of related sesquiterpenes, like (+)-a-ylangene and (-)-a-copaene.

Paper II also shows the presence of germacrene D in various plant species, i.e. in about half of the tested samples of 30 plant species (hosts and non-hosts) presented as headspace samples, essential oils and extracts. The "secondary" active compounds were only present in a few plant samples. The active compounds were isolated from essential oils by MPLC for stimulation of the RNs and for identification by GC-MS. In trying to explore the projection of the germacrene D RNs in the antennal lobe, stain was applied to the base of the sensillum from which the recordings of the germacrene D RNs were made. Out of seven experiments, one successful staining of a RN in H. assulta was obtained, showing four selectively stained axon terminals in the antennal lobe. Three of them were located close to the antennal lobe (one filling a larger volume than the others), and the fourth was located in the ventro-medial part of the lobe.

Paper III

The presence of the functionally similar germacrene D RNs in all three heliothine species raised the question whether these species also possess other types of functionally similar plant odour RN types. In H. virescens females, Røstelien et al. (2000b) identified four RN types that were co-located in the same sensilla, three of them responding to the inducible compounds E-β-ocimene, E,E-a-farnesene and E,E-4,8,12-trimethyl-1,3,7,11-tridecatetraene (TMTT), respectively. In the study reported in paper III, the three RN types were also found in females of H. armigera and H. assulta.

Additional data on the molecular receptive ranges of the three RN types in all three species were provided. The compounds activating the fourth co-located RN type were also identified in the three species, of which the flower volatile geraniol was found to be most effective. Structure-activity relationships of all four co-located RN types were worked out by the identification of the active compounds in the plant material of 19 plant species tested. The identifications were proved by retesting the authentic compounds that were commercially available. Fifty-two RNs were recorded, and the classification was based on 135 GC-SCR, which included up to 38 recordings from the single neurones.

The compounds activating each RN type were structurally similar. The RN type I was activated by the monoterpenes, E-β-ocimene, β-myrcene, Z-β-ocimene, E-4,8-dimethyl-1,3,7-nonatriene [DMNT] and dihydromyrcene, the RN type II was activated by the sesquiterpenes, E,E-a-, Z,E-a-and E-β-farnesene, the RN type III was activated by the homo-sesquiterpene TMTT and the RN type IV was activated by the monoterpene alcohols, geraniol, citronellol, (S)-(+)-and (R)-(-)-linalool and one unidentified compound. The narrowly tuning of the RNs to the few compounds was demonstrated by the presence in the plant samples of many structurally similar compounds that were inactive. The four co-located RN types in the three insect species showed the same relative response strength to the active odorants.

In trying to explore the projection of the four co-located RNs in the antennal lobe, stain was applied to the base of the sensillum from which the recordings of the RNs were made. Out of three experiments, one successful staining of the four co-located RNs in H. assulta was obtained, showing four selectively stained axons in the antennae and four axon terminals in the antennal lobe. Three of them were located in different areas close to the antennal nerve, and the fourth was located in the ventro-medial part of the lobe.

Paper IV

The presence of the numerous germacrene D RNs on the antennae of females of H. virescens (Røstelien et al. 2000a, paper II), raised the question about the biological significance of this plant odorant. A two-choice wind-tunnel was designed to study the preference of mated H. virescens females for host plants with (-)-germacrene D dispensers versus plants with control dispensers. (-)-Germacrene D (purity > 99.2%, enantiomeric purity > 99.9%), isolated from the essential oil of ylang-ylang (Cananga odorata) by MPLC, was used in the behavioural tests. The tobacco plants used in these experiments belonged to a strain that contained no germacrene D, as shown by electrophysiological experiments. Plants with dispensers having low release rates of (-)-germacrene D had a greater attractiveness to the moths than tobacco plants with control dispensers. Oviposition was observed on the plants with (-)-germacrene D, and a significantly larger number of eggs was found on these plants.

DISCUSSION

Encoding of Plant Odour Information in Receptor Neurones

The results obtained in paper I-III contribute to the understanding of the encoding of plant odour information in herbivorous insects. By the use of GC-SCR with the different GC-columns, it has been possible to identify which of the numerous plant volatiles are detected by the RNs in heliothine moths and to compare the responses within and between species. The results have shown a clear classification of the RNs in different types, each type showing a strong response to one compound and weaker responses to a few others of related structures. Since the identification is made on the basis of screening naturally produced volatiles, in principle of whole plant blends, the active compounds are probably biologically relevant, i.e. odorants for which the RNs are evolved.

This view is further strengthened by the consistent responses to the same components present in many of the plant species, and by the responses to the same compounds in the three related heliothine moths. The five types of RNs presented here constitute 91% of all olfactory RNs that were recorded. The high probability of obtaining the five RN types in stable recordings made us focus on identifying all odorants activating these RNs for determining the structure-activity relationships. In respect to the principles on how olfactory information is encoded in the RNs, the following data are important. The five RN types have narrow molecular ranges with no overlap between them, and they are as sensitive to one of the compounds as the pheromone RNs are for their key stimuli. This indicates that the information about these plant odorants is mediated to the brain according to the principle of "labelled-lines" rather than of "across fibre patterns". It means that the code to the brain of a blend quality of these odorants is the ratio of activity in the five RN types.

Narrowly Tuned Receptor Neurones

The present results of narrowly tuned RNs in heliothine moths (Paper I-III) are in agreement with results from studies of other insect species obtained by the use of GC-SCR (Wibe and Mustaparta 1996, Stensmyr et al. 2001, 2003, Barata et al. 2002, Bichão et al. 2003). The large number of plant olfactory RNs recorded from various insect species have shown narrow tuning with only limited overlap of molecular receptive ranges within chemical groups, and no overlap between different chemical groups. These data, based on GC-SCR, are more reliable than testing synthetic chemicals, since the pre-selection of the test compounds has already ignored a number of possible active odorants. Furthermore, at high concentration, one may obtain responses of the neurones without knowing what the key stimulus is. Many narrowly tuned neurones, e.g. pheromone receptor neurones, respond additionally to high concentrations of other compounds (Hansson et al. 1988). The narrow tuning of plant olfactory RNs found by GC-SCR (Paper I-III), correlates well with the results obtained in molecular biological studies, showing that one neurone expresses a single type of the identified olfactory receptor proteins (Keller and Vosshall 2003).

Interaction with a single type of receptor proteins seems more likely in neurones responding to a narrow set of related chemicals than in neurons responding to many compounds of different chemical groups. The possibility still exists in other organisms that some RNs are broadly tuned, and may perhaps have receptors interfering with certain parts of very different odour molecules (reviewed by Korsching 2002). However, data in the present study, showing reduced or no activity by small changes of the key molecule indicates the importance of the whole molecular structure for providing interaction of certain groups with the receptor protein. Since we do not yet know the molecular features of the receptor proteins, the binding characteristics between the odorant and the receptor protein can not be proposed yet.

However, the identification of genes coding for olfactory receptor proteins on the antennae of H. virescens, is in progress (Krieger et al. 2002). With more olfactory genes identified in heliothine moths and olfactory receptor proteins sequenced and characterised by crystallography, it will be interesting to study the binding that might take place between the receptor protein and the odorants identified and presented in this thesis. Furthermore, the recent identification and cloning of an OBP expressed in male (sensilla basiconica) and female (s. basiconica and s. trichodea) H. armigera (Wang et al. 2003), is promising for studying the binding of the odorants to the OBPs.

In physiological studies of vertebrate olfactory RNs, direct stimulation with a few selected commercial compounds is still common. The traditional view has been that RNs are activated by several structurally divergent molecules (i.a. Malnic et al. 1999, Duchamp-Viret et al. 1999, 2000, reviewed by Buck 2000).

One may ask whether the biological key stimulus of the olfactory RNs is present among the odorants, and whether the stimuli are presented in physiological concentrations. Despite the pre-selection of stimuli, recent studies have revealed more narrowly tuned olfactory receptors.

One example is the rat olfactory receptor (OR-I7), tested for 74 related and unrelated volatile compounds, which responded best to octanal, and weaker to heptanal and nonanal (Zhao et al. 1998). It is further shown that a difference in only one amino acid between a rat and a mouse olfactory receptor protein, change the highest affinity from octanal to heptanal, respectively (Krautwurst et al. 1998). Another study has revealed olfactory RNs with narrow tuning in general, but with overlapping molecular receptive range (Kajiya et al. 2001). The general view in vertebrates is still that the olfactory system uses a combinatorial receptor coding scheme to encode odour identities (reviewed by Buck 2000, Korsching 2002).

Obviously one expects to find broadly tuned human olfactory RNs when comparing the number of chemical stimuli humans are able to discriminate (>10 000) with the smaller number (347) of candidate functional olfactory receptor genes for the detection of odorants (Glusman et al. 2001, Zozulya et al. 2001).

Molecular Features of the Odorants

Although the receptor proteins have only recently been known, the interest for important molecular features of odorants has been shown in many studies of structure-activity relationships (i.a. Klopping 1971, Beets 1978, Ohloff 1986, Turin 1996). The structures of the odorants identified in this thesis show some significant features for stimulation of the five RN types. One important feature is the enantiomeric configuration, here shown for germacrene D (Paper I, II).

The interest for optical isomers has a long tradition in olfaction, not least since Wright (1977) proposed the hypothesis of Dyson (1938) that the interaction between the odorant and the receptors were dependent on intra-molecular vibrations (IR-spectra) of the odorant. The olfactory discrimination of optical isomers can not be explained by this hypothesis. Optical isomers, existing as mirror image molecules, have otherwise identical chemical properties, including IR-spectra. The discrimination of (+)-and (-)-carvone in humans is well known (Friedman and Miller 1971). However, we do not yet know how these enantiomers are discriminated by the receptors. The first studies, showing how information about enantiomers was received by olfactory RNs, were in bark beetles (Mustaparta et al. 1980) and in two moth species (Hansen et al. 1983), showing that a selective type of RNs had evolved for each of the enantiomers of pheromone compounds. Since then, enantioselective RNs for pheromones have been demonstrated in other insect species (reviewed by Masson and Mustaparta 1990, Leal 2001).

Obviously, since each enantiomer mediates a particular message to the insect, e.g. attraction and disruption of attraction, separate RN types have evolved for mediating the different messages of these signals. For plant odorants, it is less obvious whether the organism needs enantioselective RNs. One might expect that insects have evolved different RN types for enantiomers produced by separate enantioselective enzymes resulting in the presence of exclusively one or both isomers in various plants. Therefore, it was interesting to find that heliothine moths have only one RN type among the large number of germacrene D RNs that respond to both enantiomers but consistently with 10 times higher sensitivity for (-)-germacrene D (Paper I, II). From this, one might assume that the moths do not need to distinguish well between the two enantiomers, otherwise two RN types would have evolved.

The results on the germacrene D RN type (Paper I) are the first to show that two pure plant enantiomers activate the same RNs, but with different stimulatory effect. Another example is given in paper III, where both enantiomers of linalool elicited "secondary" responses in the geraniol RN type, (R)-(-)-linalool being more effective than the (S)-(+)-enantiomer. The fact that chiral recognition of molecules is important in biological activity (Ohloff 1994) stresses the importance of testing optically pure samples of odorants when studying the molecular receptive ranges of olfactory RNs. Obviously it is important to use chiral GC-columns for separating optical isomers, or using synthesised enantiomers of high purity in studies of molecular receptive ranges of the RNs.

In addition to chirality of the molecules, the structure-activity relationships of the five RN types (Paper I-III) also demonstrate the importance of other molecular features of the odorant in stimulating the RNs. The significance of electron dense parts and the flexibility of the molecules, reflected by the positions of double-bounds as well as the position and type of the functional groups for activating the RNs, have been shown in many previous studies (i.a. Priesner 1959, 1977, Kafka 1974, Schneider et al. 1977, Bengtsson et al. 1990, reviewed by Ohloff 1986, Masson and Mustaparta 1990, Mustaparta 2002). However, since each odorant-receptor interaction is unique, it is important to determine the molecular features in each case. In future experiments, when pure compounds of the odorants identified in the present thesis are available, more detailed quantitative structure-activity relationships (QSAR) of the five RN types can be determined in more details. This will enable the use of computational methods to point out the optimal configuration of the molecules that might be important for the interaction with the receptor proteins.

Functionally Similar Receptor Neurone Types in Related Species

Detection of the same odorants by closely related species has raised the question of whether the functional properties of the receptors are similar or different with respect to specificity. In heliothine moths, the same compounds are used as pheromones and interspecific signals, but in different compositions (reviewed by Mustaparta 1996, 1997). The previous findings of pheromone RNs with similar specificity was interesting, but perhaps not surprising considering the significance of the genetical components in the evolution of pheromone communication and the few compounds involved as signals. In contrast to this is the vast variation in volatiles released by plants.

Many insect species have evolved polyphagy whereas a few others have evolved oligophagy. Thus, one might expect diverse functional properties (specificity) of the RNs detecting plant odours. From this point of view, it was surprising to find that the identified RNs in the three heliothine species exhibit similar specificities (Paper I-III). They both detect the same key compound, and show similar sensitivity to the secondary active compounds of structural similarity. This suggests that the identified compounds play a significant role in the interaction with the plants. It is particularly interesting that the one oligophagous species, H. assulta, has evolved the same RN types as the polyphagous H. virescens and H. armigera. The results suggest that the five RN types have evolved from a common ancestor and have been unchanged in spite of employing different host plant ranges.

The conservation of the plant odour RN types in the three heltiothine species is also unexpected in considering that changes of one or a few amino acids in the sequence of a protein can result in a different compound produced in plant biosynthesis (Wang and Pichersky 1999) as well as in an olfactory receptor protein changing the affinity to a different odorant (Krautwurst et al. 1998). In the evolutionary context, the five RN types presented in this thesis seem to have the same degree of stability as most of the pheromone RN types in the heliothine moths (Mustaparta 1996, 1997), suggesting similar selection pressure for maintaining the gene types. However, different RN specificity, as found for some RNs detecting interspesific singals, is pointed out to have evolved as a consequence of other important pheromone compounds used by the species (Mustaparta 1996, 1997). It will be interesting to find out whether the remaining plant odour RN types in the heliothine moths show identical or different specificity. This will elucidate whether the use of different host plants is reflected in the RN specificity, in the CNS processing of the plant odour information or if other mechanisms are responsible for the different host range.

The Number of RN Types Responding to Plant Odorants

The identification of numerous genes coding for different types of olfactory receptor proteins in all organisms studied (Mombaerts 1999, Keller and Vosshall 2003), indicates that a large number of plant odour RN types should also be present in heliothine moths. Among the RNs recorded so far in heliothine moths, 20 different types have been identified on the basis of GC-SCR, each type responding to different compounds. In addition to the five RN types responding to compounds within the chemical groups of acyclic monoterpenes, monoterpene alcohols, sesquiterpenes and homo-terpenes (Røstelien et al. 2000a,b, Paper I-III), RN types responding to compounds of monocyclic monoterpenes, oxygenated bicyclic alcohols, aromatic hydrocarbons and "green leaf volatiles" are also recorded (Røstelien et al. unpublished results).

More RN types are expected to be found in continued GC-SCR studies of heliothine moths. This expectation is based on the finding that each RN type projects in one or a few glomeruli of the antennal lobe. In the fruitfly, 36 RN types project to the 43 glomeruli, each glomeruli receiving information from only one type (Vosshall 2001, Keller and Vosshall 2003). The number of "ordinary" glomeruli (62) in the antennal lobe of heliothine moths is assumed to be involved in plant odour information processing, which indicates a higher number than the 20 plant odour RN types identified so far (Berg et al. 2002, Skiri et al. unpublished results). In future studies it will be possible to compare the number of RN types to be identified by GC-SCR and the number of receptor protein types to be identified in molecular biological studies of heliothine moths.

The fact that the five RN types presented here constitute 91% of all of the recorded olfactory neurones responding to plant odours needs to be considered. It may reflect the relative number of each neurone type present on the antennae, assuming that all neurones are equally accessible to the electrophysiological recordings. This suggests that the five RN types are of particular importance for the heliothine species. Alternatively, the other neurones may belong to sensilla that are not frequently penetrated because of the localisation. Although we have never found indications of RN types with particular localisation, we can not exclude this possibility. The next question is in which glomeruli the different RN types, project.

A functional organisation of glomeruli has been shown for the MGC in heliothine moths (Berg et al. 1998, Berg 1998). Since the plant odour RNs are localised in the shorter s. trichodea type II and s. basiconica, it has not yet been possible to carry out recordings from cut hairs with the "tip-recordings" technique. Therefore, we explored the possibility of staining the RNs selectively by applying the stain to the recording site of the tungsten microelectrode, after making sure that the neurone was penetrated by the electrode (observed by the rise in spike activity followed by silence). The preliminary results showing selective stained axons terminating in 4 glomeruli (Paper II and III), are promising for further studies of the functional organisation of the 62 "ordinary" glomeruli. Interestingly, optical imaging experiments reveal calcium responses in corresponding areas of the antennal lobe of H. virescens when stimulating with the same plant odorants (Skiri et al. 2002, unpublished results). Of particular interest is the projection of the (-)-germacrene D RNs that is easily obtained in heliothine antennae. The projections of the RNs in heliothine moths seem to have no branching in more glomeruli (Paper II and III) in contrast to single primary olfactory axons of the locust projecting in several glomeruli (Laurent and Naraghi 1994, Anton and Hansson 1996).

Behavioural Responses to Plant Odorants

In the identification of plant odorants that are detected by the oligophagous H. assulta and the polyphagous H. armigera and H. virescens, the interesting question is which odorants and mechanisms make these moths select different host plants. If all RN types are functionally similar, the different behavioural response to the compounds must be based on different processing of the odour information in the CNS. This is shown for the insect produced signals in some heliothine species, being detected by functionally similar RNs, but eliciting different behaviours (Mustaparta 1996, 1997). Alternatively, the different heliothine moths may have species specific RN types for some plant odours, which are not yet identified. In order to resolve these questions, more behavioural studies are needed.

In the present thesis, only behavioural responses to (-)-germacrene D were examined (Paper IV). Since attraction to plants is thought to be elicited by blends rather than single compounds, (-)-germacrene D was added to host plants containing no germacrene D. Dispensers with a low release rate of (-)-germacrene D placed on the host plant, caused increased attraction and oviposition of mated H. virescens females to the plant compared with the control plant. Whether the increased number of eggs laid was an indirect result from the increased attraction or (-)-germacrene D also had an effect on oviposition was not elucidated. Another interesting question is whether the related sesquiterpenes activating the (-)-germacrene D RNs elicit the same behavioural response at higher doses. This would be expected if the response is only dependent on increased activation in these RNs, and the compounds are ineffective on other RNs. Of particular interest is to test the behavioural effect of (+)-germacrene D, for comparing with the response to (-)-germacrene D. In field studies in Australia, attraction to (-)-germacrene D by H. armigera females is under investigation (Gregg and Del Socorro, unpublished results).

Among the other plant odorants identified in the present thesis (Paper I-III), myrcene, Z-β-ocimene and racemic linalool are in previous studies found attractive for H. armigera females when presented as constituents of synthetic blends (Rembold et al. 1991, Bruce and Cork 2001). We can only speculate about the behavioural role of the other identified odorants (Paper I-III). Geraniol is a typical floral scent. A plant in the flowering phase provides the food source, and in the floral and fruit phases the plants are in the most attractive phases for ovipositing heliothine moths (reviewed by Fitt 1989). The feeding of H. virescens caterpillars is shown to induce certain volatiles in host plants, which repel conspecific, mated females (De Moraes et al. 2001). Thus, the odorants activating the three other co-located RN types may elicit avoidance reaction, since these compounds are induced in several host plants upon attack by feeding larvae (reviewed by Turlings and Benrey 1998, Paré and Tumlinson 1999, Dicke and Van Loon 2000).

Learning of Plant Odorants

The studies of this thesis have focused on which odorants of host and non-host plants are detected by the heliothine moths. The extremely sensitive RNs, particularly for the key stimuli (nanograms for (-)-germacrene D), indicate that the odorants may be detected at some distance from the plant, making the insect able to decide whether to fly toward the plant or not. In this situation, one may assume that the quality and the ratio of the odorants are important for the attraction to the host plant. When arriving on the plant, the concentration of the odorants might be quite high, and also the odorants with weak effect might be detected. After landing, taste and mechano-sensation are more important for the further evaluation of the plant as a host for nutrition or egg-laying (Ramaswamy 1988, Ramaswamy et al. 1987, Jackson et al. 1984, Li et al. 2001).

When exposed to the relatively high concentration of the whole blend of the flower during nectar feeding, the insects learn to associate the odour with the nutritional reward, which is also reflected by increased probability of selecting the same host species after experience (Firempong and Zalucki 1991, Cunningham et al. 1998a,b). In an ongoing project, the ability of heliothine moths to learn the plant odorants identified (Paper I-III) is studied by the use of the PER (Skiri et al., unpublished results). Interesting questions about olfactory discriminatin can be answerd by these time consuming but rather simple experiments; do the heliothine moths show different ability of learning plant odorants used in different contexts, and do they manage to discriminate information about odorants that are encoded in the same RNs? In our laboratorium, neuronal connections between the olfactory and taste pathways in H. virescens females are also studied.

Projections in the suboesophageal ganglion of taste RNs on the antennae and proboscis have been described (Jørgensen 2003, Kvello 2003), and intracellular recordings combined with stainings have shown one neurone with similar morphology to the VUM neurone in the honeybee (Rø et al. unpublished results). The results in the present thesis contribute to these projects on olfactory coding and learning, by providing knowledge about biologically relevant plant odorants. Altogether the aim is to elucidate the neuronal mechanisms underlying olfactory coding and olfactory learning involved in nectar feeding as well as in host plant selection by heliothine moths.

CONCLUSIONS AND PROSPECTS

The results of this study show that plant odour information in females of the three heliothine moths H. virescens, H. armigera and H. assulta, is received by narrowly tuned RNs with non-overlapping molecular receptive ranges. The five RNs described are all strongly activated by one key odorant and show weaker response to a few related molecules. The RN types are functionally similar within and between the three heliothine species. A new staining method is tested which show projections of the RNs in the antennal lobe. One odorant, being the key stimulus for the major RN type, was found to be attractive when added to host plants that did not contain this odorant. The identified biologically relevant odorants provide a basis for further studies of olfaction in heliothine moths. This knowledge is presently used in studies of olfactory coding and olfactory learning in addition to behavioural experiments. In future experiments combining physiology, molecular biology and chemistry, the identified odorants may be utilised to study the molecular and cellular function of olfactory RNs.

4

Importance of Bryophytes in Plant Kingdom

INTRODUCTION

The term bryophyta means a moss plant, derived from the Greek words bryon meaning moss and phyton meaning plant. It was Robert Brown who first introduced this term in 1864 to include the algae, the fungi, the lichens and the mosses. However, in recent years this term has been used to denote this group of plants which includes the Hepaticae, the Anthocerotae and the Musci.

Bryophytes occupy an important place in the plant kingdom, although they were thought to be of little economic value. They are the simplest and the most primitive of the land plants. They are the pioneers to colonize terrestrial habitat from aquatic environment. Their adaptation to a terrestrial mode of life is partial as water is indispensable in one stage or another in their life cycle. Hence, they are also known as the amphibians of the plant kingdom. They have a remarkable capacity to absorb water and turn fresh in no time, which has given them the name the resurrection plants. These plants occur more commonly during rainy seasons and also in humid areas. Their growth is gregarious and within a short period they become so abundant as to occupy large areas, but usually do not form a very conspicuous part of the vegetation. Though basically terrestrial, there are a few aquatic forms such as Riccia fluitans, Ricciocarpus natans and Riella spp.

Moreover Cryptothallus and Buxbaumia are saprophytic genera of liverworts and mosses on contrary to the rest of the bryophytes, which are autotrophic. Usually, bryophytes show a preference to inhabit microclimatic niches such as crevices of rocks and trees, near small shady springs and so on. But they are seen growing on a wide range of substrata. They may be old discarded/abandoned leather, rubber and wooden goods, tiled and asbestos roofs, on mortar of stone and mud walls. Besides these, growing as epiphytes on barks of trees (corticolous), on leaves (epiphyllous or folicolous), on rocks (rupicolous), on pebbles and stones (saxicolous), on fallen logs (lignicolous), river banks and road side cuts is a common sight. A notable feature of certain bryophytes is their remarkable capacity to endure drought and are usually known as the Xeromorphic forms. They form the secondary colonizers after lichens on barren rocks.

THE IMPORTANCE OF BRYOPHYTES

Bryophytes are useful to human beings as well as to other organisms and the environment in many ways. They are the secondary colonizers on barren rocks next to lichens in plant succession in xerosere. Thus they help in weathering of soil. They are extremely good soil binders as they form large mats on forest floors and roadside cuts, thus controlling soil erosion. They are a good source of humus and hence a haven for a number of soil-dwelling invertebrates like earthworms. They form very good seed beds for seedlings and saplings particularly in evergreen forests. They are very good indicators of environmental pollution. Thallus extracts of Marchantia polymorpha and Lunularia cruciata show antifungal and antibacterial properties. A number of liverworts are known to be rich sources of terpenoids such as sesqui-and diterpenoids, and also aromatic compounds. They are very good laboratory specimens to carry out experiments on hybridization, apogamy, apospory, DNA targeting etc. and also to understand the process of evolution of land plants. They are also used in decorations in bouquets, vases etc.

Bryophytes are not only used by man but also by other vertebrates and invertebrates. Birds gathering bryophytes particularly mosses has already been reported by Pant and Tewari (1981). Whether to decorate their nests or to camouflage it is a little hard to explain. Similarly, bagworms and some spiders also gather bryophytes for their nests to camouflage themselves from predators.

THE DIVERSITY

So far, 960 genera and about 20,000 species of bryophytes are known to Plant Science. India is one of the 12 megabiodiversity countries in the world. The large area and the variety of phytoclimatic conditions met within its different biogeographical zones contribute to the great diversity of the Indian flora. These biogeographical zones with diverse topographical variations and climatic conditions, have been divided into seven bryogeographical regions. They are the Western Himalayas, the Eastern Himalayas, the Punjab and the West Rajasthan, the Gangetic Plains, the Central India, the Western and the Eastern Ghats and the Deccan Plateau. Recently, the islands of Andaman and Nicobar in the Bay of Bengal have been proposed as the eighth region to be included.

The Western Ghats also known as the Sahyadris in peninsular India is identified as one of the 'hotspots' in the world. The flora of this region has been studied with emphasis on the flowering plants and even the pteridophytes. However, our knowledge on the taxonomy and distribution of bryophytes is far from adequate and still relies on the work done during the late 19th and early 20th centuries. So far about 850 species of liverworts belonging to 140 genera and 52 families and 2000 species of mosses belonging to 342 genera and 54 families are reported to occur in India. From the West Coast and the Western Ghats 121 species of liverworts with 10 endemic and 682 species of mosses with 190 endemic have been reported.

Authors of the plant names are abbreviated following Brummit and Powell (1992) and titles of journals and books are based on B-(otanico)-P-(eriodicum)-H-(untianum) (Lawrence & al., 1968), B-P-H/S (Bridson & Smith,1991), Taxonomic literature ed. 2 (Stafleu & Eowan, 1976-1986) and TL2 Supplements (Stafleu & Mennega, 1992-2000) respectively.

EARLIER WORK

The earliest work on Indian mosses dates back to 1741 by Dillenius in his Historia muscorum. This was followed much later by Linnaeus in 1753, where he included some Indian mosses along with higher plants in his Species Plantarum. Later, Buchanan Hamilton, a medical officer in the British Embassy in 1802, made a collection of mosses from Nepal which was published by Hooker in 1808 as Musci Nepalensis. It was only after this more serious studies on mosses were taken up by Hooker (1818-1820) where he described several new species. In 1825 he jointly made two publications with Greville on Indian mosses. In 1826-27, Schwägrichen included some Indian mosses in his supplement to Hedwig's Species Muscorum.

Our present knowledge of bryophytes relies mostly on such early works as Wallich (1828-32), Harvey and Hooker (1840), Griffith (1842, 1843). The first bryophyte from the Southern Western Ghats was reported by Van Rheedes (1678-1703) in his monumental work on South Indian plants, Hortus Indicus Malabaricus. He described with illustrations one moss from Kerala as 'poovan-peda' (Vol. 12. t. 37. p. 71. 1693). This was later identified by Nicolson & al. (1988) as Bryum bicolor Dickson, in contradiction to the earlier interpretation as B. coronatum Schwägr. Montagne (1842) reported 100 species of mosses from Nilgris (Tamil Nadu) in his Cryptogamae Nilgheriensis, Müller (1853) reported over 100 species of mosses from Nilgris in his Musci Neilgherrenses, Mitten (1859) reported over 700 species in his Musci Indiae Orientalis. Subsequently, Brotherus (1899) reported 86 species and Dixon (1909, 1911, 1912, 1914, 1921, 1938). From a few notable parts of the Western Ghats, Potier de la Varde (1928) described 100 species of mosses from the tropical rain forests of Silent Valley in Kerala, (1922-24) from the Western Ghats of Madurai (Tamil Nadu) in his Musci Madurensis, Foreau (1961, 1964) listed 368 species from the Palni hills (W. Ghats of Madurai) which included 95 new species and 15 varieties, and Raghavan and Wadhwa (1968, 1970) from Agumbe and Hulical ranges in Shimoga district of Karnataka.

The only other comprehensive work on Indian mosses that gives some information on the habitat and distribution of the south Indian mosses including the Western Ghats is Gangulee's Mosses of eastern India and adjacent regions (1969-1980). More recent works are Mohamed's study (1981, 1982, 1984) on

Rhodobryum commersonii (Schwägr.) Paris and R. aubertii (Schwägr.) Thér. in South India, Bryum wightii Mitt. and related species, and a synopsis of the genus Rhodobryum in Asia respectively. Ochi (1985) published an annotated list of mosses of the subfamily Bryoideae in south, southeast and east Asia, Ellis (1989) made a revision on the genus Calymperes in southern India and neighbouring islands and Crum (1972) published a taxonomic account of the family Erpodiaceae in which there is a mention of one collection from the Travancore hills.

Similarly, the present knowledge of the liverworts relies mostly on the works of Mitten (1861) who reported 290 species of liverworts in his Hepaticae Indiae Orientalis, Stephani (1900-1924), Benedix (1953), Kashyap (1929, 1932), Chopra (1938, 1943), Bharadwaj (1950, 1958, 1960), Kachroo (1969, 1970a, 1970b, 1973), Kachroo & al. (1977), Udar (1976), Udar and Srivastava (1975, 1977), Joshi and Biradar (1984), Asthana and Srivastava (1991), Nath and Asthana (1998), Asthana & al. (1995), which pertain to different localities in southern India. The most recent works are by Daniels 1998, 2001, Daniels & Daniel 2003a, 2003b, 2004, Kumar and Maniselvan (1994), Maniselvan and Kumar (1998), Maniselvan and Kumar (2000a, 2000b), Nair and Madhusoodhanan (2001, 2002) in Kerala and that of Easa (2003) which is a documentation of bryophytes of Kerala, mostly based on earlier works.

Name of the Species-Systematic Position Mosses

Aerobryidium aureonitens (Hook. ex Schwägr.) Broth Meteoriaceae

Aerobryidium filamentosum (Hook.) M.Fleisch. Meteoriaceae

Aerobryidium wallichii (Brid.) C.C.Towns. Meteoriaceae

Aerobryopsis longissima (Dozy & Molk.) M.Fleisch. Meteoriaceae

Aerobryopsis membranacea (Mitt.) Broth. Meteoriaceae

Anacamptodon validinervis Dixon & P. de la Varde Fabroniaceae

Anoectangium aestivum (Hedw.) Mitt. Pottiaceae

Anoectangium stracheyanum Mitt. Pottiaceae

Anomobryum auratum (Mitt.) A.Jaeger Bryaceae

Anomobryum julaceum (Schrad. ex P.Gaertn., B.Mey. & Scherb.) Schimp. Bryaceae

Anomobryum schmidii (Müll.Hal.) A.Jaeger Bryaceae

Anomobryum subnitidum Cardot & P. de la Varde Bryaceae

Aptychella speciosa (Mitt.) Tixier Sematophyllaceae

Aptychella tenuiramea (Mitt.) Tixier Sematophyllaceae

Archidium birmannicum Mitt. ex Dixon Archidiaceae

Archidium microthecium Dixon & P. de la Varde Archidiaceae

Archidium ohioense Schimp. ex Müll.Hal. Archidiaceae

Astomum minutum Dixon P. de la Varde Pottiaceae

Atrichum longifolium Cardot & Dixon ex Gangulee Polytrichaceae

Atrichum pallidum Renauld & Cardot Polytrichaceae

Barbella pendula (Sull.) M. Fleisch. Meteoriaceae

Barbellopsis trichophora (Mont.) W.R. Buck Meteoriaceae

Barbula dharvarensis Dixon Pottiaceae

Barbula indica (Hook.) Spreng. Pottiaceae

Barbula javanica Dozy & Molk. Pottiaceae

Barbula spathulifolia (Dixon & P. de la Varde) R.H. Zander Pottiaceae

Barbula vardei Pottiaceae

Bartramia gathica Cardot & P. de la Varde Bartramiaceae

Bartramia madurensis Dixon & P. de la Varde Bartramiaceae

Bartramidula roylei (Hook.f.) Bruch & Schimp. Bartramiaceae

Bellibarbula recurva (Griff.) R.H. Zander Pottiaceae

Brachymenium acuminatum Harv. Bryaceae

Brachymenium bryoides Hook. ex Schwägr. Bryaceae

Brachymenium capitulatum (Mitt.) Paris Bryaceae

Brachymenium exile (Dozy & Molk.) Bosch. & Sande Lac. Bryaceae

Brachymenium indicum (Dozy & Molk.) Bosch. & Sande Lac. Bryaceae

Brachymenium leptophyllum (Bruch & Schimp. ex Müll.Hal.) Bruch & Schimp.

Bryaceae

Brachymenium nepalense Hook. Bryaceae

Brachymenium pendulum Mont. Bryaceae

Brachymenium systylium (Müll.Hal.) A.Jaeger Bryaceae

Brachymenium walkeri Broth. Bryaceae

Brachythecium buchananii (Hook.) A.Jaeger Brachytheciaceae

Brachythecium nitidissimum Dixon & P. de la Varde Brachytheciaceae

Brachythecium plumosum (Hedw.) Bruch, Schimp. & W.Gümbel Brachytheciaceae

Brachythecium procumbens (Mitt.) A.Jaeger Brachytheciaceae

Brachythecium salebrosum (F.Weber & D.Mohr) Bruch, Schimp. & W.Gümbel

Brachytheciaceae

Braunia apiculata Cardot Hedwigiaceae

Braunia macrocarpa (Müll.Hal.) A.Jaeger Hedwigiaceae

Braunia secunda Hook. Hedwigiaceae

Breutelia dicranacea (Müll.Hal.) Mitt. Bartramiaceae

Breutelia microdonta (Mitt.) Broth. Bartramiaceae

Brothera leana (Sull.) Müll.Hal. Dicranaceae

Bryhnia decurvans (Mitt.) Dixon Brachytheciaceae

Bryocrunia vivicolor (Broth. & Dixon) W.R. Buck Sematophyllaceae

Bryoerythrophyllum inaequalifolium (Taylor) R.H. Zander Pottiaceae

Bryoerythrophyllum wallichii (Mitt.) P.C. Chen Pottiaceae

Bryum alpinum Huds. ex With. Bryaceae

Bryum apalodictyoides Müll.Hal. Bryaceae

Bryum apiculatum Schwägr. Bryaceae

Bryum argenteum Hedw. var. argenteum Bryaceae

Bryum argenteum Hedw. var. australe Rehmann ex Dixon Bryaceae

Bryum aubertii (Schwägr.) Brid. Bryaceae

Bryum billardierei Schwägr. Bryaceae

Bryum caespiticium Hedw. Bryaceae

Bryum capillare Hedw. Bryaceae

Bryum cellulare Hook. Bryaceae

Bryum coronatum Schwägr. Bryaceae

Bryum euryphyllum Cardot & P. de la Varde Bryaceae

Bryum lamprostegrum Müll.Hal. Bryaceae

Bryum neelgheriense Mont. var. neelgheriense Bryaceae

Bryum pachycladum Cardot ex P. de la Varde Bryaceae

Bryum pallescens Schleich ex Schwägr. Bryaceae

Bryum paradoxum Schwägr. Bryaceae

Bryum pseudotriquetrum (Hedw.) P.Gaertn. var. pseudotriquetrum Bryaceae

Bryum retusifolium Cardot & P. de la Varde Bryaceae

Bryum thomsonii Mitt. Bryaceae

Bryum uliginosum (Brid.) Bruch & Schimp. Bryaceae

Bryum wightii Mitt. Bryaceae

Calymperes afzelii Sw. Calymperaceae

Calymperes erosum Müll.Hal. Calymperaceae

Calymperes graeffeanum Müll.Hal. Calymperaceae

Calymperes lonchophyllum Schwägr. subsp. lonchophyllum Calymperaceae

Calymperes motleyi Mitt. ex Dozy & Molk. Calymperaceae

Calymperes tenerum Müll.Hal. Calymperaceae

Calyptothecium auriculatum (Dixon) Nog. Neckeraceae

Calyptothecium oxyphyllum Dixon & P. de la Varde Neckeraceae

Calyptothecium recurvulum (Broth.) Broth. Neckeraceae

Calyptothecium symphysodontoides Dixon & P. de la Varde Neckeraceae

Calyptothecium wightii (Mitt.) M. Fleisch. Neckeraceae

Campylodontium perplicatum (Thér. & P. de la Varde) Broth. Entodontaceae

Campylopodium phascoides (Müll.Hal.) Paris Dicranaceae

Campylopus andreanus Cardot & P. de la Varde Dicranaceae

Campylopus comosus (Reinw. & Hornsch.) Bosch & Sande Lac. Dicranaceae

Campylopus ericoides (Griff.) A.Jaeger Dicranaceae

Campylopus flagelliferus (Müll.Hal.) A.Jaeger Dicranaceae

Campylopus gracilis (Mitt.) A.Jaeger Dicranaceae

Campylopus introflexus (Hedw.) Brid. var. introflexus Dicranaceae

Campylopus involutus (Müll.Hal.) A.Jaeger Dicranaceae

Campylopus nivalis (Brid.) Brid.var. nivalis Dicranaceae

Campylopus pilifer Brid. Dicranaceae

Campylopus richardii Brid. Dicranaceae

Campylopus savannarum (Müll.Hal.) Mitt. Dicranaceae

Campylopus schmidii (Müll.Hal.) A.Jaeger Dicranaceae

Campylopus subfragilis Renauld & Cardot Dicranaceae

Campylopus umbellatus (Arn.) Paris Dicranaceae

Campylopus zollingerianus (Müll.Hal.) Bosch & Sande Lac. Dicranaceae

Ceratodon purpureus (Hedw.) Brid. var. purpureus Ditrichaceae

Ceratodon purpureus (Hedw.) Brid. var. stenocarpus Ditrichaceae

Chaetomitrium papillifolium Bosch & Sande Lac. Hookeriaceae

Chenia leptophylla (Müll.Hal.) R.H. Zander Pottiaceae

Chrysocladium retrorsum (Mitt.) M. Fleisch. Meteoriaceae

Claopodium assurgens (Sull. & Lesq.) Cardot Thuidiaceae

Claopodium pellucinerve (Mitt.) Best Thuidiaceae

Claopodium prionophyllum (Müll.Hal.) Broth. Thuidiaceae

Cryptopapillaria fuscescens (Hook.) M. Menzel Meteoriaceae

Ctenidium lychnites (Mitt.) Broth. Hypnaceae

Diaphanodon blandus (Harv.) Renauld & Cardot Trachypodaceae

Dicranella divaricata (Mitt.) A.Jaeger Dicranaceae

Dicranella griffithii (Mitt.) Giese & J.-P.Frahm Dicranaceae

Dicranella heteromalla Hedw. Dicranaceae

Dicranella stricticaulis Cardot & P. de la Varde Dicranaceae

Dicranodontium denudatum (Brid.) E.Britton Dicranaceae

Dicranodontium uncinatum (Harv.) A.Jaeger Dicranaceae

Dicranoloma brevisetum (Dozy & Molk.) Paris Dicranaceae

Dicranum dilatinerve Cardot & P. de la Varde Dicranaceae

Dicranum psathyrum Klazenga Dicranaceae

Didymodon rigidulus Hedw. Pottiaceae

Diphyscium fasciculatum Mitt. Diphysciaceae

Diphyscium mucronifolium Mitt. Diphysciaceae

Distichophyllum montagneanum Hookeriaceae

Distichophyllum succulentum (Mitt.) Broth. Hookeriaceae

Ditrichum amoenum (Thwaites & Mitt.) Paris Ditrichaceae

Ditrichum difficile (Duby) M.Fleisch. Ditrichaceae

Ditrichum tortipes (Mitt.) Kuntze var. strictum Dixon & P. de la Varde Ditrichaceae

Ditrichum tortipes (Mitt.) Kuntze var. tortipes Ditrichaceae

Ectropothecium andrei Cardot & P. de la Varde Hypnaceae

Ectropothecium cyperoides (Hook.) A. Jaeger Hypnaceae

Ectropothecium densum Dixon & P. dela Varde Hypnaceae

Ectropothecium drepanocladioides Broth. & P. de la Varde Hypnaceae

Ectropothecium laevigatum Thwaites & Mitt. Hypnaceae

Ectropothecium manii Broth. Hypnaceae

Ectropothecium zollingeri Sematophyllaceae

Entodon macropodus (Hedw.) Müll.Hal. Entodontaceae

Entodon myurus (Hook.) Hampe Entodontaceae

Entodon obtusatus Broth. Entodontaceae

Entodontopsis nitens (Mitt.) W.R. Buck Plagiotheciaceae

Entodontopsis wightii (Mitt.) W.R. Buck Plagiotheciaceae

Entosthodon diversinervis Müll.Hal. Funariaceae

Entosthodon perrottetii (Mont.) Müll.Hal. Funariaceae

Entosthodon physcomitrioides (Mont.) Mitt. Funariaceae

Entosthodon pilifer Mitt. Funariaceae

Entosthodon planifolius Thwaites & Mitt. Funariaceae

Entosthodon submarginatus Müll.Hal. Funariaceae

Entosthodon wichurae M.Fleisch. Funariaceae

Erpodium biseriatum (Austin) Austin Erpodiaceae

Erpodium mangiferae Müll.Hal. Erpodiaceae

Erythrodontium julaceum (Hook. ex Schwägr.) Paris Entodontaceae

Eurhynchium vagans (A.Jaeger) E.B.Bartram Brachytheciaceae

Fabronia curvirostris Dozy & Molk. Fabroniaceae

Fabronia goughii Mont. Fabroniaceae

Fabronia madurensis Dixon & P. de la Varde Fabroniaceae

Fabronia pusilla Raddi Fabroniaceae

Fabronia schmidii Müll.Hal. Fabroniaceae

Fabronia secunda Mont. Fabroniaceae

Fissidens amplifolius Dixon & P. de la Varde Fissidentaceae

Fissidens anomalus Mont. Fissidentaceae

Fissidens asplenioides Hedw. Fissidentaceae

Fissidens auriculatus Müll.Hal. Fissidentaceae

Fissidens beckettii Mitt. Fissidentaceae

Fissidens bryoides Hedw. Fissidentaceae

Fissidens ceylonensis Dozy & Molk. var. acutifoilus Dixon & P. de la Varde

Fissidentaceae

Fissidens ceylonensis Dozy & Molk. var. ceylonensis Fissidentaceae

Fissidens crenulatus Mitt. var. crenulatus Fissidentaceae

Fissidens crispulus Brid. var. crispulus Fissidentaceae

Fissidens crispulus Brid. var. robinsonii (Broth.) Z.Iwats. & Z.H.Li Fissidentaceae

Fissidens diversifolius Mitt. Fissidentaceae

Fissidens dubius P.Beauv. Fissidentaceae

Fissidens excedens Broth. Fissidentaceae

Fissidens flaccidus Brid. Fissidentaceae

Fissidens ganguleei Nork. Fissidentaceae

Fissidens gardneri Mitt. Fissidentaceae

Fissidens grandifrons Brid. Fissidentaceae

Fissidens griffithii Gangulee Fissidentaceae

Fissidens hyalinus Hook.f. & Wilson Fissidentaceae

Fissidens incognitus Gangulee Fissidentaceae

Fissidens involutus Mitt. subsp. curvato-involutus (Dixon) Gangulee Fissidentaceae

Fissidens involutus Mitt. subsp. involutus Fissidentaceae

Fissidens jungermannioides Griff. Fissidentaceae

Fissidens kalimpongensis Gangulee Fissidentaceae

Fissidens karwarensis Dixon Fissidentaceae

Fissidens microdictyon Dixon & P. de la Varde Fissidentaceae

Fissidens perumalensis Dixon & P. de la Varde Fissidentaceae

Fissidens schmidii Müll.Hal. Fissidentaceae

Fissidens serratus Müll.Hal.var. serratus Fissidentaceae

Fissidens subangustus M.Fleisch. Fissidentaceae

Fissidens subbryoides Gangulee Fissidentaceae

Fissidens subpulchellus Nork. Fissidentaceae

Fissidens teraicola Müll.Hal. Fissidentaceae

Fissidens virens Thwaites & Mitt. Fissidentaceae

Fissidens walkeri Broth. Fissidentaceae

Fissidens zippelianus Dozy & Molk. Fissidentaceae

Fissidens zollingeri Mont. Fissidentaceae

Floribundaria floribunda (Dozy & Molk.) M. Fleisch. Meteoriaceae

Floribundaria thuidioides M. Fleisch. Meteoriaceae

Foreauella orthothecia (Schwägr.) Dixon & P. de la Varde Sematophyllaceae

Forsstroemia indica (Mont.) Paris Cryphareaceae

Funaria excurrentinervis Cardot & P. de la Varde Funariaceae

Funaria hygrometrica Hedw. var. calvescens (Schwägr.) Mont. Funariaceae

Funaria hygrometrica Hedw. var. hygrometrica Funariaceae

Funaria pulchra Dixon & P. de la Varde Funariaceae

Funaria sinuato-limbata Cardot & P. de la Varde Funariaceae

Gamiella ceylonensis (Broth. in Herzog) B.C. Tan & W.R. Buck Sematophyllaceae

Garckea flexuosa (Griff.) Margad. & Nork. Ditrichaceae

Glossadelphus anisopterus (Cardot & P. de la Varde) Broth. Sematophyllaceae

Glossadelphus ivoreanus (Mitt.) M. Fleisch. Sematophyllaceae

Grimmia indica (Dixon & P. de la Varde) Goffinet & Greven Grimmiaceae

Grimmia ovalis (Hedw.) Lindb. Grimmiaceae

Gymnostomiella vernicosa (Hook.) M. Fleisch. Pottiaceae

Gymnostomum calcareum Nees & Hornsch. Pottiaceae

Haplocladium microphyllum subsp. virginianum (Brid.) Reimers Thuidiaceae

Hedwigidium integrifolium (P.Beauv.) Dixon Hedwigiaceae

Herpetineuron toccoae (Sull. & Lesq.) Cardot Thuidiaceae

Himantocladium cyclophyllum (Mitt.) M. Fleisch. Neckeraceae

Holomitrium densifolium (Wilson) Wijk. & Margad. Dicranaceae

Homalia trichomanoides (Hedw.) Schimp. Neckeraceae

Homaliadelphus targionianus (Mitt.) Dixon & P. de la Varde Neckeraceae

Homaliodendron exiguum (Bosch & Sande Lac.) M. Fleisch. Neckeraceae

Homaliodendron flabellatum (Sm.) M. Fleisch. Neckeraceae

Homaliodendron microdendron (Mont.) M. Fleisch. Neckeraceae

Homaliodendron montagneanum (Müll.Hal.) M. Fleisch. Neckeraceae

Homalothecium sericeum Schimp. Brachytheciaceae

Hookeria acutifolia Hook. & Grev. Hookeriaceae

Hyophila comosa Dixon & P. de la Varde Pottiaceae

Hyophila grandiretis Dixon & P. de la Varde Pottiaceae

Hyophila involuta (Hook.) A. Jaeger Pottiaceae

Hyophila mollifolia Dixon & P. de la Varde Pottiaceae

Hyophila spathulata (Harv.) A. Jaeger Pottiaceae

Hyophila viridula Cardot & P. de la Varde Pottiaceae

Hypnum cupressiforme Hedw. Hypnaceae

Hypopterygium tamarisci (Sw.) Brid. ex Müll.Hal. Hypopterygiaceae

Isopterygium albescens (Hook.) A. Jaeger Hypnaceae

Isopterygium lignicola (Mitt.) A. Jaeger Hypnaceae

Isopterygium undulatum Dixon & P. de la Varde Hypnaceae

Isotheciopsis comes (Griff.) Nog. Meteoriaceae

Jaegerina stolonifera Müll.Hal. var. incrassata P. de la Varde Pterobryaceae

Jaegerina stolonifera Müll.Hal. var. stolonifera Pterobryaceae

Juratzkaea indica Broth. & P. de la Varde Fabroniaceae

Lepidopilidium furcatum (Thwaites) Mitt. Hookeriaceae

Leptotrichella denticulate (Cardot & P. de la Varde) Ochyra Ditrichaceae

Leptotrichella schmidii (Müll.Hal.) Ochyra Ditrichaceae

Leskea consanguinea (Mont.) Mitt. Leskeaceae

Leucobryum aduncum Dozy & Molk Leucobryaceae

Leucobryum aduncum Dozy & Molk. var. scalare (Müll.Hal.) A. Eddy Leucobryaceae

Leucobryum bowringii Mitt. Leucobryaceae

Leucobryum humillimum Cardot Leucobryaceae

Leucobryum javense (Brid.) Mitt. Leucobryaceae

Leucobryum juniperoideum (Brid.) Müll.Hal. Leucobryaceae

Leucoloma amoene-virens Mitt. Dicranaceae

Leucoloma molle (Müll.Hal.) Mitt. Dicranaceae

Leucoloma nitens (Thwaites & Mitt.) A.Jaeger Dicranaceae

Leucoloma taylorii (Schwägr.) Mitt. Dicranaceae

Leucoloma tenerum Mitt. Dicranaceae

Leucophanes glaucum (Schwägr.) Mitt. Leucobryaceae

Leucophanes octoblepharioides Brid. Leucobryaceae

Levierella fabroniacea Müll.Hal. Fabroniaceae

Lopidium struthiopteris (Brid.) M. Fleisch. Hypopterygiaceae

Macgregorella indica (Broth.) W.R.Buck Fabroniaceae

Macrocoma tenuis subsp. sullivantii (Müll.Hal.) Vitt Orthotrichaceae

Macromitrium calymperoideum Mitt. Othotrichaceae

Macromitrium japonicum Dozy & Molk. Orthotrichaceae

Macromitrium moorcroftii (Hook. & Grev.) Schwägr. Orthotrichaceae

Macromitrium nepalense (Hook. & Grev.) Schwägr. Orthotrichaceae

Macromitrium polygonostomum Dixon & P. de la Varde Orthotrichaceae

Macromitrium schmidii Müll.Hal. Orthotrichaceae

Macromitrium sulcatum (Hook.) Brid. Orthotrichaceae

Macromitrium turgidum Dixon Orthotrichaceae

Macrothamniella pilosula (Mitt.) M.Fleisch. Hylocomiaceae

Macrothamnium macrocarpum (Reinw. & Hornsch.) M.Fleisch. Hylocomiaceae

Mesonodon flavescens (Hook.) W.R.Buck Entodontaceae

Meteoriopsis reclinata (Müll.Hal.) M. Fleisch. Meteoriaceae

Meteoriopsis squarrosa (Hook.) M. Fleisch. Meteoriaceae

Meteorium buchananii (Brid.) Broth. Meteoriaceae

Meteorium polytrichum Dozy & Molk. Meteoriaceae

Meteorium subpolytrichum subsp. subpolytrichum Meteoriaceae

Microcampylopus khasianus (Griff.) Giese & J.-P.Frahm Dicranaceae

Mielichhoferia schmidii Müll.Hal. Bryaceae

Nanothecium foreaui Dixon & P. de la Varde Entodontaceae

Neckera aequalifolia Müll.Hal. Neckeraceae

Neckera andrei Thér. & P. de la Varde Neckeraceae

Neckera goughiana Mitt. Neckeraceae

Neckera pennata Hedw. var. rhytidiodonta Dixon & P. de la Varde Neckeraceae

Neckera pennata Hedw. Neckeraceae

Neckera semicrispa Cardot & P. de la Varde Neckeraceae

Neckeropsis exserta (Hook. ex Schwägr.) Broth. Neckeraceae

Noguchiodendron sphaerocarpum (Nog.) Ninh & Pocs Neckeraceae

Octoblepharum albidum Hedw. Leucobryaceae

Oedicladium rufescens (Reinw. & Hornsch.) Mitt. Myuriaceae

Oreoweisia laxifolia (Hook.f.) Kindb. Dicranaceae

Oxyrrhynchium ovatum Cardot & P. de la Varde Brachytheciaceae

Palamocladium leskeoides (Hook.) E.Britton Brachytheciaceae

Papillaria crocea (Hampe) A. Jaeger Meteoriaceae

Papillaria leuconeura (Müll.Hal.) A. Jaeger Meteoriaceae

Pelekium contortulum (Mitt.) Touw Thuidiaceae

Pelekium investe (Mitt.) Touw Thuidiaceae

Pelekium versicolor (Hornsch. ex Müll.Hal.)Touw Thuidiaceae

Philonotis dispersa (Cardot & P. de la Varde) D.G.Griffin & W.R.Buck Bartramiaceae

Philonotis falcata (Hook.) Mitt. Bartramiaceae

Philonotis fontana (Hedw.) Brid. Bartramiaceae

Philonotis hastata (Duby) Wijk. & Margad. Bartramiaceae

Philonotis mollis (Dozy & Molk.) Mitt. Bartramiaceae

Philonotis subrigida Cardot & P. de la Varde Bartramiaceae

Philonotis thwaitesii Mitt. Bartramiaceae

Philonotis tomentella Molendo Bartramiaceae

Physcomitrium coorgense Broth. Funariaceae

Pinnatella alopecuroides (Hook.) M. Fleisch. Neckeraceae

Pinnatella foreauana Thér. & P. de la Varde Neckeraceae

Plagiomnium integrum (Bosch & Sande Lac.) T.J. Kop. Mniaceae

Plagiomnium rhynchophorum (Hook.) T.J. Kop. Mniaceae

Plagiomnium rostratum (Schrad.) T.J. Kop. Mniaceae

Plagiomnium succulentum (Mitt.) T.J. Kop. Mniaceae

Plagiothecium neckeroideum var. madurense Dixon & P. de la Varde Plagiotheciaceae

Plagiothecium vesiculariopsis Dixon & P. de la Varde Plagiotheciaceae

Plagithecium neckeroideum var. sikkimense Renauld & Cardot Plagiotheciaceae

Platydictya madurensis (Cardot & P. de la Varde) R.S.Chopra Amblystegiaceae

Platyhypnidium muelleri (A.Jaeger) M.Fleisch. Amblystegiaceae

Platyhypnidium riparioides (Hedw.) Dixon Amblystegiaceae

Pleuridium denticulatum (Müll.Hal.) Mitt. Ditrichaceae

Pogonatum aloides (Hedw.) P. Beauv. Polytrichaceae

Pogonatum inflexum (Lindb.) Sande Lac. Polytrichaceae

Pogonatum microstomum (Schwägr.) Brid. Polytrichaceae

Pogonatum neesii (Müll.Hal.) Dozy Polytrichaceae

Pogonatum patulum (Harv.) Mitt. Polytrichaceae

Pogonatum perichaetiale (Mont.) A. Jaeger subsp. perichaetiale Polytrichaceae

Pogonatum subtortile (Müll.Hal.) A. Jaeger Polytrichaceae

Pogonatum urnigerum (Hedw.) P. Beauv. Polytrichaceae

Pohlia camptotrachela (Renauld & Cardot) Broth. Bryaceae

Pohlia elongata Hedw. Bryaceae

Pohlia flexuosa Hook. Bryaceae

Pohlia trematodontea (Müll.Hal.) Broth. Bryaceae

Pseudoleskeopsis zippelii (Dozy & Molk.) Leskeaceae

Pseudosymblepharis bombayensis (Müll.Hal.) P. Sollman Pottiaceae

Pseudotaxiphyllum distichaceum (Mitt.) Z. Iwats. Hypnaceae

Pseudotaxiphyllum pohliaecarpum (Sull. & Lesq.) Z. Iwats. Hypnaceae

Pterobryopsis acuminata (Hook.) M. Fleisch. Pterobryaceae

Pterobryopsis divergens (Mitt.) Nog. Pterobryaceae

Pterobryopsis gedehensis M. Fleisch Pterobryaceae

Pterobryopsis orientālis (Müll.Hal.) M. Fleisch. Pterobryaceae

Pterobryopsis schmidii (Müll.Hal.) M. Fleisch. Pterobryaceae

Pterobryopsis scrabriuscula (Mitt.) M. Fleisch. Pterobryaceae

Ptychomitrium tortula (Harv.) A. Jaeger Ptychomitriaceae

Pylaisia falcata Schimp. Hypnaceae

Pylaisiadelpha capillacea (Griff.) B.C. Tan & Y. Zia Sematophyllaceae

Pyrrhobryum spiniforme (Hedw.) Mitt. Rhizogoniaceae

Racomitrium subsecundum (Hook. & Grev.) Mitt. var. brachyphyllum Cardot & P. de la

Varde Grimmiaceae

Racopilum cuspidigerum (Schwägr.) Ångstr. Racopilaceae

Racopilum orthocarpum Wilson ex Mitt. Racopilaceae

Racopilum schmidii (Müll.Hal.) Mitt. var. breviastrum Cardot Racopilaceae

Racopilum schmidii (Müll.Hal.) Mitt. var. schmidii Racopilaceae

Radulina hamata (Dozy & Molk.) W.R. Buck & B.C. Tan Sematophyllaceae

Regmatodon orthostegius Mont. Leskeaceae

Rhachithecium perpusillum (Thwaites & Mitt.) Broth. Orthotrichaceae

Rhamphidium madurense Dixon & P. de la Varde Ditrichaceae

Rhaphidorrhynchium cucullifolium (Cardot & Dixon) Broth. Sematophyllaceae

Rhaphidorrhynchium sebillei Broth. & Thér. Sematophyllaceae

Rhaphidostichium subleptocarpum (Thér. & P. de la Varde) Broth. Sematophyllaceae

Rhodobryum commersonii (Schwägr.) Paris Bryaceae

Rhodobryum giganteum (Schwägr.) Paris Bryaceae

Rhodobryum roseum (Hedw.) Limpr. Bryaceae

Rhynchostegiella humillima (Mitt.) Broth. Brachytheciaceae

Rhynchostegiella leiopoda Dixon & Cardot Brachytheciaceae

Rhynchostegium brachythecioides Dixon & P. de la Varde Brachytheciaceae

Rhynchostegium celebicum (Sande Lac.) A.Jaeger Brachytheciaceae

Rhynchostegium javanicum (Bél.) Besch. Brachytheciaceae

Schlotheimia grevilleana Mitt. Orthotrichaceae

Schoenobryum concavifolium (Griff.) Gangulee Cryphaeaceae

Schwetschkea applanata (Thwaites & Mitt.) Broth. Fabroniaceae

Schwetschkeopsis elongata (Dixon & P. de la Varde) W.R.Buck & H.A.Crum

Fabroniace

Sematophyllum humile (Mitt.) Broth. Sematophyllaceae

Sematophyllum subhumile (Müll.Hal.) M. Fleisch. Sematophyllaceae

Sematophyllum subpinnatum (Brid.) E. Britton Sematophyllaceae

Sphaerothecium reconditum Thwaites & Mitt. Dicranaceae

Stereophyllum radiculosum (Hook.) Mitt. Plagiotheciaceae

Symphyodon complanatus Dixon Neckeraceae

Symphyodon pygmaeus (Broth.) S. He & Snider Neckeraceae

Symphysodontella involuta (Thwaites & Mott.) M. Fleisch. Pterobryaceae

Syntrichia fragilis (Taylor) Ochyra Pottiaceae

Syrrhopodon gardneri (Hook.) Schwägr. Calymperaceae

Syrrhopodon prolifer Schwägr. var. prolifer Calymperaceae

Syrrhopodon semiliberus (Mitt.) Besch. ex Paris Calymperaceae

Taxiphyllum isopterygioides (Dixon) W.R.Buck Sematophyllaceae

Taxiphyllum minutirameum (Müll.Hal.) H.A.Mill. & D.R.Sm. Hypnaceae

Taxiphyllum taxirameum (Mitt.) M.Fleisch. Hypnaceae

Thamniopsis utacamundiana (Mont.) W.R.Buck Hookeriaceae

Thamnobryum alleghaniense (Müll.Hal.) Nieuwl. Neckeraceae

Thamnobryum alopecurum (Hedw.) Nieuwl. ex Gangulee Neckeraceae

Thamnobryum parvulum (Mitt.) R.S. Chopra Neckeraceae

Thamnobryum suserratum (Hook.) Nog. & Z. Iwats. Neckeraceae

Thuidium cymbifolium (Dozy & Molk.) Dozy & Molk. Thuidiaceae

Thuidium plumulosum (Dozy & Molk.) Dozy & Molk. Thuidiaceae

Thuidium pristocalyx var. pristocalyx (Müll.Hal.) A.Jaeger Thuidiaceae

Timmia megapolitana Hedw. Pottiaceae

Timmiella anomala (Bruch, Schimp. & W.Gümbel) Limpr. Pottiaceae

Toloxis semitorta (Müll.Hal.) W.R.Buck Meteoriaceae

Trachycladiella sparsa (Mitt.) M.Menzel Meteoriaceae

Trachyloma indicum var. indicum Mitt. Pterobryaceae

Trachyphyllum inflexum (Harv.) A.Gepp. Entodontaceae

Trachypodopsis serrulata (P.Beauv.) M.Fleisch. var. crispatula (Hook.) Zanten

Trachypodaceae

Trachypus bicolor Reinw. & Hornsch. var. bicolor Trachypodaceae

Trachypus bicolor var. hispidus (Müll.Hal.) Cardot Trachypodaceae

Trachypus humilis Lindb. var. humilis Trachypodaceae

Trachypus humilis Lindb. var. tenerrimus (Broth. ex Herzog) Zanten Trachypodaceae

Trematodon conformis Mitt. Dicranaceae

Trematodon longicollis Michx. Dicranaceae

Trematodon schmidii Müll.Hal. Dicranaceae

Trichostomum hyalinoblastum (Broth.) Broth. Pottiaceae

Trichostomum tenuirostre (Hook. & Taylor) Lindb. Pottiaceae

Ulota schmidii (Müll.Hal.) Mitt. Orthotrichaceae

Vesicularia firma Dixon & P. de la Varde Hypnaceae

Vesicularia nitidula Cardot & P. de la Varde Hypnaceae

Vesicularia reticulata (Dozy & Molk.) Broth. Hypnaceae

Vesicularia subpilicuspis Cardot & P. de la Varde Hypnaceae

Warburgiella leptorhynchoides (Mont.) M. Fleisch Sematophyllaceae

Warburgiella perviridis Dixon & P. de la Varde Sematophyllaceae

Weissia controversa Hedw. Pottiaceae

Weissia edentula Mitt. Pottiaceae

Weissia macrocarpa Cardot & P. de la Varde Pottiaceae

Wilsoniella decipiens (Mitt.) Alston var. acutifolia (Dixon) Wijk. & Margad.

Ditrichaceae

Zygodon acutifolius Müll.Hal. Orthotrichaceae

Zygodon humilis Thwaites & Mitt. Orthotrichaceae

Zygodon reinwardtii (Hornsch.) A.Braun Orthotrichaceae

Zygodon tetragonostomus A.Braun ex Bruch, Schimp. & W.Gümbel Orthotrichaceae

Liverworts

Acrolejeunea fertilis (Reinw. & al.) Schiffn. Lejeuneaceae

Aneura pellioides (Horik.) Inoue Aneuraceae

Aneura pinguis (L.) Dumort. Aneuraceae

Archilejeunea polymorpha (Sande Lac.) B.Thiers & Gradst. Lejeuneaceae

Asterella blumeana (Nees) Kachroo Aytoniaceae

Asterella leptophylla (Mont.) Pandé & al. Aytoniaceae

Asterella vulcanica (Schiffn.) Pandé & al. Aytoniaceae

Asterella wallichiana (Lehm. & Lindenb.) Aytoniaceae

Calypogeia arguta Mont. & Nees Calypogeiaceae

Cheilolejeunea birmensis (Steph.) Mizut. Lejeuneaceae

Cheilolejeunea imbricata (Nees) S.Hatt. Lejeuneaceae

Cheilolejeunea intertexta (Lindenb.) Steph. Lejeuneaceae

Cheilolejeunea khasiana (Mitt.) N.Kitag. Lejeuneaceae

Cheilolejeunea laeviuscula (Mitt.) Steph. Lejeuneaceae

Cheilolejeunea serpentina (Mitt.) Mizut. Lejeuneaceae

Cheilolejeunea udarii G.Asthana & al. Lejeuneaceae

Chonecolea schusteri Udar & A.Kumar Chonecoleaceae

Cololejeunea furcilobulata (Berrie & E.W.Jones) R.M.Schust. Lejeuneaceae

Cololejeunea lanciloba Steph. Lejeuneaceae

Cololejeunea minutissima (Sm.) Schiffn. Lejeuneaceae

Cololejeunea spinosa (Horik.) S.Hatt. Lejeuneaceae

Cyathodium aureonitens (Griff.) Mitt. Targioniaceae

Cyathodium smaragdinum Schiffn. Targioniaceae

Cyathodium tuberosum Kashyap Targioniaceae

Drepanolejeunea angustifolia (Mitt.) Grolle Lejeuneaceae

Drepanolejeunea ternatensis (Gottsche) Steph. Lejeuneaceae

Dumortiera hirsuta (Sw.) Nees Marchantiaceae

Exormotheca ceylonensis Meijer Exormothecaceae

Exormotheca tuberifera Kashyap Exormothecaceae

Fossombronia crispata Lindb. Fossombroniaceae

Fossombronia foreaui Udar & S.C.Srivast. Fossombroniaceae

Fossombronia himalayensis Fossombroniaceae

Frullania acutiloba Mitt. Frullaniaceae

Frullania apiculāta (Reinw. & al.) Dumort. Frullaniaceae

Frullania arecae Mitt. Frullaniaceae

Frullania campanulata Sande Lac. Frullaniaceae

Frullania densiloba Steph. Frullaniaceae

Frullania ericoides (Nees) Mont. Frullaniaceae

Frullania muscicola Steph. Frullaniaceae

Frullania neurota Taylor Frullaniaceae

Frullania serrata Gottsche Frullaniaceae

Frullania tamarisci (L.) Dumort. Frullaniaceae

Herbertus dicranus (Taylor ex Gottsche & al.) Trevis. Herbertaceae

Herbertus pinnatus (Steph.) H.A.Mill Herbertaceae

Heteroscyphus argutus (Reinw. & al.) Schiffn. Geocalycaceae

Heteroscyphus orbiculatus A.Srivast. & S.C.Srivast. Geocalycaceae

Heteroscyphus palniensis A.Srivast. & S.C.Srivast. Geocalycaceae

Heteroscyphus perfoliatus (Mont.) Schiffn. Geocalycaceae

Jungermannia tetragona Lindenb. Jungermanniaceae

Lejeunea neelgherriana Gottsche Lejeuneaceae

Leptolejeunea balansae Steph. Lejeuneaceae

Leptolejeunea elliptica (Lehm. & Lindenb.) Schiffn. Lejeuneaceae

Leptolejeunea himalayensis Pandé & R.N.Misra Lejeuneaceae

Leptolejeunea maculata (Mitt.) Schiffn. Lejeuneaceae

Leptolejeunea sikkimensis Udar & U.S.Awasthi Lejeuneaceae

Leucolejeunea xanthocarpa (Lehm. & Lindenb.) A.Evans Lejeuneaceae

Lophocolea bidentata (L.) Dumort. Geocalycaceae

Lophocolea cuspidate (Nees) Limpr. Geocalycaceae

Lophocolea heterophylla (Schrad.) Dumort. Geocalycaceae

Lophocolea muricata (Lehm.) Schiffn. Geocalycaceae

Lopholejeunea subfusca (Nees) Schiffn. Lejeuneaceae

Lunularia cruciata (L.) Dumort. Lunulariaceae

Mannia foreaui Udar & V.Chandra Aytoniaceae

Marchantia indica Kashyap Marchantiaceae

Marchantia nepalensis Lehm. & Lindenb. Marchantiaceae

Marchantia palmata Nees Marchantiaceae

Marchantia polymorpha L. Marchantiaceae

Mastigolejeunea auriculata (Wilson & Hook.) Schiffn. Lejeuneaceae

Metzgeria conjugata Lindenb. Metzgeriaceae

Metzgeria consanguinea Schiffn. Metzgeriaceae

Metzgeria furcata (L.) Dumort. Metzgeriaceae

Metzgeria himalayensis Kashyap Metzgeriaceae

Metzgeria indica Udar & S.C.Srivast. Metzgeriaceae

Metzgeria lutescens Steph. Metzgeriaceae

Metzgeria nilgiriensis Udar & S.C.Srivast. Metzgeriaceae

Metzgeria pandei S.C.Srivast. & Udar Metzgeriaceae

Microlejeunea ulicina A.Evans Lejeuneaceae

Notoscyphus darjeelingensis Udar & A.Kumar Jungermanniaceae

Notoscyphus lutescens (Lehm.) Mitt. Jungermanniaceae

Notoscyphus pandei Udar & A.Kumar Jungermanniaceae

Notoscyphus paroicus Schiffn. Jungermanniaceae

Pallavicinia lyellii (Hook.) S.Gray Pallaviciniaceae

Petalophyllum indicum Kashyap Fossombroniaceae

Plagiochasma appendiculatum Lehm. & Lindenb. Aytoniaceae

Plagiochasma japonicum (Steph.) C.Massal. Aytoniaceae

Plagiochasma pterospermum C.Massal. Aytoniaceae

Plagiochasma rupestre (J.R.Forst. & G. Forst.) Steph. Aytoniaceae

Plagiochila elegans Mitt. Plagiochilaceae

Plagiochila indica Mitt. ex Steph. Plagiochilaceae

Plagiochila peradenyensis Schiffn. Plagiochilaceae

Porella campylophylla (Lehm. & Lindb.) Trevis. subsp. campylophylla Porellaceae

Porella campylophylla (Lehm. & Lindb.) Trevis. subsp. lancistipula (Steph.) S.Hatt.

Porellaceae

Porella perottetiana (Mont.) Trevis. var. perottetiana Porellaceae

Radula madagascariensis Gottsche Radulaceae

Radula nilgiriensis Udar & A.Kumar Radulaceae

Radula onstraedii Radulaceae

Reboulia hemisphaerica (L.) Raddi Aytoniaceae

Riccardia levierii Schiffn. Riccardiaceae

Riccardia multifida (L.) S.Gray Riccardiaceae

Riccardia personii Srivast. & Udar Riccardiaceae

Riccardia tenuicostata Schiffn. Riccardiaceae

Riccia aravalliensis Pandé & Udar Ricciaceae

Riccia billardieri Mont. & Nees Ricciaceae

Riccia crozalsii Levier Ricciaceae

Riccia discolor Lehm. & Lindenb. Ricciaceae

Riccia fluitans L. Ricciaceae

Riccia gangetica Ahmad Ricciaceae

Riccia grollei Udar Ricciaceae

Riccia melanospora Kashyap Ricciaceae

Riccia poihaiana A.E.D.Daniels & P.Daniel Ricciaceae

Riccia sorocarpa Bisch. Ricciaceae

Riccia velimalaiana A.E.D.Daniels & P.Daniel Ricciaceae

Riccia warnstorfii Limpr. Ricciaceae

Schiffneriolejeunea indica (Steph.) Udar & R.N.Misra Lejeuneaceae

Spruceanthus semirepandus Nees Lejeuneaceae

Targionia hypophylla L. Targioniaceae

Hornworts

Anthoceros bharadwajii Udar & Asthana Anthocerotaceae

Anthoceros crispulus (Mont.) Douin Anthocerotaceae

Anthoceros erectus Kashyap Anthocerotaceae

Anthoceros macrosporus Steph. Anthocerotaceae

Anthoceros punctatus L. Anthocerotaceae

Notothylas dissecta Steph. Notothylaceae

Notothylas indica Kashyap Notothylaceae

Phaeoceros carolianus (Michx.) Prosk. Anthocerotaceae

Phaeoceros laevis Prosk. Anthocerotaceae

ENDEMIC/ENDANGERED BRYOPHYTES OF TAMILNADU

As no CAMP (Conservation Assessment Management Plan) has been done for any group of bryophytes in India, the exact status of these plants is not known. Therefore, the following list has been prepared mostly from my own surveys and 10 years of field experience and also based on the little information provided by Udar & Srivastava (1983), Vohra & Aziz (1997) and Singh (1997). However, due to the fast vanishing forests which leads to habitat loss, pollution and global warming, almost all bryophytes are gradually becoming vulnerable, endangered and some in a state of extinction in Tamilnadu. All hornworts are highly endangered in Tamilnadu.

MOSSES

Anacamptodon validinervis

Archidium microthecium

Bartramia madurensis

Dicranum dilatinerve

Distichophyllum succulentum

Ditrichum tortipes var. strictum

Ectropothecium densum
Entosthodon diversinervis
Fabronia curvirostris
Fabronia madurensis
Fabronia schmidii
Fissidens perumalensis
Glossadelphus anisopterus
Juratzkaea indica
Mielichoferia schmidii
Nanothecium foreaui
Neckera aequalifolia
Neckera goughiana
Neckera semicrispa
Oxyrrhynchium ovatum
Philonotis subrigida
Pinnatella foreauana
Plagiothecium neckeroideum var. madurense
Plagiothecium vesiculariopsis
Pleuridium denticulatum
Pohlia trematodontea
Racopilum schmidii var. breviastrum
Racopilum schmidii var. schmidii
Rhaphidorrhynchium sebillei
Rhynchostegiella leiopoda
Rhynchostegium brachythecioides
Schwetschkea indica
Sematophyllum subhumile
Thamnobryum parvulum
Trematodon schmidii
Trichostomum hyalinoblastum
Vesicularia nitidula

Vesicularia subpilicuspis
Weissia macrocarpa
Zygodon acutifolius

LIVERWORTS

Fossombronia foreaui
Cheilolejeunea udarii
Heteroscyphus orbiculatus
Heteroscyphus palniensis
Heteroscyphus perfoliatus
Lophocolea bidentata
Lophocolea cuspidata
Lophocolea heterophylla
Lophocolea muricata
Metzgeria indica
Metzgeria nilgiriensis
Metzgeria pandei
Radula nilgiriensis
Mannia foreaui
Riccia grollei
Riccia poihaiana
Riccia warnstorfii
Riccia velimalaiana

RARE BRYOPHYTES OF TAMILNADU

MOSSES

Aerobryopsis membranacea
Anoectangium stracheyanum
Aptychella tenuiramea
Archidium birmannicum
Bellibarbula recurva
Brachymenium indicum

Brachymenium walkeri
Brachythecium nitidissimum
Brachythecium procumbens
Braunia secunda
Breutilia dicranacea
Brothera leana
Bryum euryphyllum
Bryum neelgheriense var. neelgheriense
Calymperes motleyi
Calyptothecium oxyphyllum
Calyptothecium symphysodontoides
Ectropothecium andrei
Ectropothecium cyperoides
Ectropothecium drepanocladioides
Ectropothecium laevigatum
Ectropothecium manii
Ectropothecium zollingeri
Fissidens crispulus var. robinsonii
Fissidens griffithii
Fissidens hyalinus
Fissidens jungermannioides
Fissidens karwarensis
Fissidens microdictyon
Fissidens subbryoides
Fissidens subpulchellus
Foreauella orthothecia
Forsstroemia indica
Funaria excurrentinervis
Funaria hygrometrica Hedw. var. calvescens
Funaria hygrometrica Hedw. var. hygrometrica
Funaria pulchra

Funaria sinuato-limbata
Hypopterygium tamarisci
Jaegerina stolonifera var. incrassata
Jaegerina stolonifera var. stolonifera
Physcomitrium coorgense
Pinnatella alopecuroides
Plagiomnium integrum
Pterobryopsis divergens
Regmatodon orthostegus
Sematophyllum humile
Ulota schmidii
Warburgiella perviridis

LIVERWORTS

Aneura pellioides
Aneura pinguis
Asterella blumeana
Aterella leptophylla
Asterella vulcanica
Asterella wallichiana
Calypogeia arguta
Chonecolea schusteri
Cololejeunea spinosa
Drepanolejeunea angustifolia
Drepanolejeunea ternatensis
Dumortiera hirsuta
Fossombronia crispata
Frullania neurota
Frullania serrata
Herbertus dicranus
Herbertus pinnatus
Lejeunea neelgherriana

Leptolejeunea balansae
Marchantia indica
Marchantia nepalensis
Marchantia palmata
Marchantia polymorpha
Notoscyphus darjeelingensis
Plagiochasma appendiculatum
Plagiochasma japonicum
Plagiochasma pterospermum
Radula madagascariensis
Radula onstraedii
Reboulia hemisphaerica
Targionia hypophylla

HORNWORTS

Anthoceros bharadwajii
Anthoceros crispulus
Anthoceros erectus
Anthoceros macrosporus
Anthoceros punctatus
Notothylas dissecta
Notothylas indica
Phaeoceros carolianus
Phaeoceros laevis

SUGGESTIONS

1. There is an urgent need to carryout a systematic floristic study on the bryophytes of South India particularly the Western and the Eastern Ghats.
2. Identify areas with luxuriant growth of bryophytes and try to understand their ecology.
3. Minimize environmental pollution and interference in the forests.

4. A habitat approach of conservation is mandatory.
5. However, if this is not possible the only other alternative is ex situ conservation by developing bryophyte gardens in glass and/or green houses.
6. Finally, make an in vitro propagation of threatened and endangered species to be re-introduced into the wild.

MAJOR GAPS IN TAMILNADU

- Lack of experts in the field
- Only a few educational and research institutions with good laboratory facilities
- No national or statelevel website or database available
- Lack of availability of literature
- The rate at which taxonomic revisions are being done outside India is not incorporated in India
- Lack of awareness towards this group of plants
- No reference herbarium available
- Very few sporadic studies and no systematic study carried out
- Lack of financial support to carry out surveys and taxonomic studies on this group

5

Weed Identification and Control

COMMON WEED IDENTIFICATION TERMS

Alternate-leaves that are arranged singly up the stem; not opposite each other. Annual-plant that germinates, flowers, seeds, and dies during one growing season. Anther-structure in a flower in which pollen is formed Auricle-lobe-like structure at the collar of a grass leaf. Awn-slender bristle at the tip of grass seed structures. Axil-the angle formed between a leaf and a stem. Basal-at the base of a plant or plant part. Bi-ennial-plant that germinates in one growing season, then flowers, seeds, and dies during the next year.

Bract-leaf-like structure at the base of flowers or leaves.

Calyx-all the flower leaves together, normally green in color.

Clasping leaves-leaves that appear to wrap around the stem at their base.

Compound leaves-leaves with 2 or more distinct leaflets.

Cotyledons-the first leaf-like structures that appear after germination; seed leaves.

Crown-the structure formed where leaves, stems, and roots grow together.

Dissected-deeply and repeatedly divided into smaller parts.

Entire-not toothed or otherwise cut.

Glumes-the 2 bracts surrounding a grass spikelet.

Head-a group of flowers borne tightly together.

Leaflets-leaf-like structures within a compound leaf.

Ligule-the structure at the collar of a grass leaf between the sheath and the stem.

Linear-long, narrow, and slender.

Lobed-a cut into a leaf from the edge toward the center; greater than toothed, but not quite compound.

Margin-the edge of a leaf.

Membranous-thin and flexible, usually not green.

Midrib-the center and usually most prominent vein on a leaf.

Nodding-a flower that is not pointed upward, but bent downward or sidewise to the stem.

Opposite-leaves situated directly across the stem from each other.

Ovate-egg shaped in outline.

Panicle-a much-branched inflorescence.

Perennial-a plant that lives for more than 2 growing seasons.

Petiole-a leaf stalk.

Pinnate-with 2 rows of leaflets, like a feather.

Plume-a hair-like or feather-like structure, often on a seed.

Pubescence-the hairs on a leaf, stem or flower.

Rhizome-a creeping, underground stem.

Rosette-a circular, normally basal, clump of leaves.

Sheath-the extension of leaf tissue surrounding a stem.

Simple leaf-one with a blade in one piece; not compound.

Spike-a narrow, non-spreading inflorescence.

Spikelet-a single or group of floral structures in a grass.

Spur-a hollow appendage on a flower.

Stolon-a creeping stem along the surface of the ground.

Succulent-fleshy and juicy.

Taproot-a thick, central root with minimal branching.

Trifoliate leaf-a leaf made of 3 leaflets; clover-like.

Whorled-3 or more similar structured arranged as spokes on a wheel.

A weed is any plant growing in an area where it is not wanted. We try to control weeds because they compete with crops for light, moisture, space and nutrients. Certain weed species can harbor plant diseases and insect pests and can be a serious threat to the ecosystem's health. Other species may be poisonous, allergenic or irritant to humans and/or livestock. Medical and economic problems such as illness, death, rash, hayfever, or a reduction in quality of fur, meat and milk products may result.

Weeds have many unique characteristics which make them extremely difficult to control. Most produce a tremendous number of seeds. The seeds of some weed species may be dormant for many years, with only a small percentage germinating each year. Some seeds will not germinate without intense heat, like from a wild fire. Weeds generally mature earlier than the crop and often seeds will be dropped before crop harvest and remain in the field. Weeds are generally more competitive than crops and can often survive under unfavorable growing conditions.

Weeds may conveniently be divided into two classes based on the way in which they emerge from the seed. Monocots emerge with a single seed leaf whereas dicots emerge with two seed leaves. Most monocot weeds found in turfgrass are from the family Gramineae and are termed weedy grasses. Examples include crabgrass, annual bluegrass, tall fescue, and quackgrass. Dicots, on the other hand, are termed broadleaf weeds and include such plants as dandelion, clover, ground ivy, knotweed, and plantain.

Weedy grasses and broadleaf weeds are further divided into groups according to the plants' length of life. Perennial weeds have a life of more than two years, though new seeds may be produced every year. Biennial weeds have a life of two years, generally storing up food reserves in the leaves and roots the first year and producing seed in the second year. The biennial weeds are often grouped with perennial weeds since control is similar.

Annual weeds germinate from seed, grow, flower, and produce seed in less than one year. Summer annuals (AKA warm season annuals) germinate in the spring and mature in the fall, whereas winter annuals (AKA cool season annuals) germinate in fall or late winter and mature in late spring.

Summer annuals complete their lifecycle from seed to maturity in less than one year. They germinate in the spring, mature, set seed and die in the fall. Winter annuals germinate in the fall, overwinter as seedlings or small rosettes and mature, set seed and die the following spring or early summer. Some weeds are capable of both summer and winter annual lifecycles.

Biennials

Biennials complete their lifecycles in less than two years. Germination and the production of an overwintering rosette of leaves occur the first year. The second year flowering, seed production, and plant death occur. Control is best obtained during the first year.

Perennials

Perennials live for more than two years. They reproduce vegetatively from roots, rhizomes, buds, or tillers, or from seed, or both. They can be especially difficult to control because of their persistent root systems.

Early identification of emerged weed species is critical for choosing the best weed control methods. This guide will enable you to identify weeds at three growth stages. Effective control of weeds in turf is based on correct identification. Many books and charts are available to help in identifying common lawn weeds. For additional help in weed identification, inquire at your county extension service.

What is a Weed?

Generally, the term weed is used to describe any plant that is unwanted and grows or spreads aggressively. Terms such as invasive, exotic or non-native are used somewhat interchangeably to refer to weeds that infest large areas. Free from the natural controls present in their native lands, these weeds grow quickly and overtake native plants.

Noxious Weed

Millions of acres of once healthy, productive rangelands, forestlands and riparian areas have been overrun by noxious or invasive weeds. They are invading recreation areas, public lands, National Parks, State Parks, roadsides, streambanks, Federal, state, and private lands.

Invasive or noxious weeds:

- o destroy wildlife habitat
- o reduce opportunities for hunting, fishing, camping and other recreational activities
- o displace many Threatened and Endangered Species
- o reduce plant and animal diversity because of weed monocultures-single plat species that over run all others in an area
- o disrupt waterfowl and neo-tropical migratory bird flight patterns and nesting habitats
- o cost millions of dollars in treatment and loss of productivity to private land owners.

What is a Noxious Weed?

The term "weed" means different things to different people. In the broadest sense, it is any plant growing where it is not wanted. Weeds can be native or non-native, invasive or non invasive, and noxious or not noxious. Legally, a noxious weed is any plant designated by a Federal, State or county government as injurious to public health, agriculture, recreation, wildlife or property. (Sheley, Petroff, and Borman,1999) A noxious weed is also commonly defined as a plant that grows out of place (i.e. a rose can be a weed in a wheat field) and is "competitive, persistent, and pernicious." (James, et al, 1991).

The noxious weeds mandated for control are plants non-native to North America. Consequently, these plants do not have the natural checks as found in their native land, such as insects, diseases, and herbivores that would keep the plant population in check. Due to the competitive aggressive ability of these plants coupled with no natural controls, these plants will develop mono-culture stands. Not only are many crops out competed by these weeds but native vegetation and the wildlife associated with it will be replaced. Consequently, identifying the weeds when they first become established and developing an integrated weed management plan to control them is critical in maintaining healthy, productive land.

The term noxious weed is used to describe a legal designation for plant species that have been determined to be especially undesirable or difficult to control. These weeds are subject, by law, to certain restrictions. Regulated by the U.S. Department of Agriculture, there are 90 federal noxious weeds. Please see the Federal Listing of Noxious and Alien weeds in the rear of this course for more information.

Are Invasive Plants the Same as Noxious Weeds?

No. Invasive plants include not only noxious weeds, but also other plants that are not native to this country. Plants are considered invasive if they have been introduced into an environment where they did not evolve. As a result, they usually have no natural enemies to limit their reproduction and spread (Westbrooks, 1998). Some invasive plants can produce significant changes to vegetation, composition, structure, or ecosystem function. (Cronk and Fuller, 1995).

What is an Invasive Species?

An 'invasive species' is defined as a species that is 1) non-native (or alien) to the ecosystem under consideration and 2) whose introduction causes or is likely to cause economic or environmental harm or harm to human health.

Weed Classifications

Most common weeds fit into two large general classifications, broadleaves and grasses. Broadleaves and grasses may be further divided into annuals and perennials. Annual and perennial weeds may be even further subdivided by the seasons in which they germinate and grow.

ALFALFA

Hops Clover Trifolium dubium

Trifolium dubium (formerly T. minus) goes by several common names, including Least Hops Clover, Small Hops

Clover, Lesser Trefoil, Yellow Shamrock, & Kleiner Klee. The tiny flowers are said to resemble h Hops Clover hops.

It is a very common mat-forming weed which erupts on roadsides & empty lots early in spring in recently disturbed soil or wherever grass, including wild meadow grasses, were chopped short at the end of the previous year. It is only two to six inches tall, so tries to get a headstart on larger plants, germinating its previous year's seeds in January or February.

It can be the dominant weed or miniature wildflower in April. Though it may be still be blooming May through July, other plants will have grown so much as to hide it, & the early-spring mats will be overwhelmed by larger weeds & grasses. But it will still be seen until at least early July popping up in cleared gardening areas.

It is easily mistaken for another common wildflower, Black Medick (Medicago lupulina, but Black Medick's wee yellow flowers turn into a cluster of tiny kidney-shaped black seeds, & its trifoil leaves have pointy tips. Black Medick's range is world-wide, & Least Hops Clover is only slightly less universal in its distribution. Both plants became distributed throughout the world at least a century ago, when sailing ships routinely picked up alfalfa bales from Eurasia & Central Asia to use as passing fill-in between more profitable shipments, then sold the bales to farmers in ports of the New World & Pacific Islands. Sundry knapweeds also spread throughout the world from this practice.

Like other clovers & legume family plants generally, Hops Clover helps bind nitrogen in poor soils. Though some people fight it as an unwanted weed, it is in reality doing the soil a good deed. Though adaptable to a wide range of conditions, it tends to die out of rich soils, & does not compete well with big perennials. But for the most part, it cannot be gotten rid of, so it might as well be appreciated for the real charm of so many teensy bright yellow flowers & miniature trifoil leaves. While Hops Clover annoys people who want perfection in their lawns, others never even notice it is all over the place. It is too small to make a particularly big display. It is one of the smallest of clovers, though even at that, the tiny flowers are actually made up of about two-dozen even more wee florets.

COMMON WEEDS

Field Bindweed Convolvulus arvensis Russian Knapweed Acroptilon repens.

Field bindweed can be spread by seed, root fragments, farm implements, infested soil adhering to the roots of nursery stock, root growth from infested areas, and by animals. Field bindweed has a deep root system that competes with crop plants for water and nutrients. Vines climb on plants and shade crops, cause lodging of small grains, and make harvesting difficult by clogging machinery. Dense field bindweed infestations may reduce crop yields by 50 to 60 percent. Land infested with field bindweed is reduced in value.

Field bindweed is a long-lived perennial which produces a dense ground cover. The twining stems vary from 1.5 to 6 feet or more in length. Leaf size and shape are variable, but generally the leaves are 1 to 2 inches long, smooth and shaped like an arrowhead. Flowers are funnelshaped, about 1 inch diameter, and white or pink in color. The flower stalk has two small bracts located ½ to 2 inches below the flower. The bracts, along with leaf shape and smaller flower size, distinguish field bindweed from hedge bindweed.

Russian knapweed (Acroptilon repens) is a creeping, herbaceous perennial of foreign origin that reproduces from seed and vegetative root buds. Shoots, or stems, are erect, 18 to 36 inches tall, with many branches. Lower leaves are 2 to 4 inches long and deeply lobed. Upper leaves are smaller, generally with smooth margins, but can be slightly lobed.

Shoots and leaves are covered with dense gray hairs. The solitary, urn-shaped flower heads occur on shoot tips and generally are 1/4 to 1/2 inch in diameter with smooth papery bracts. Flowers can be pink, lavender or white. Russian knapweed has vertical and horizontal roots that have a brown to black, scaly appearance, especially apparent near the crown. Russian knapweed emerges in early spring, bolts in May to June, and flowers through the summer into fall. Russian knapweed is toxic to horses.

The key to Russian knapweed control is to stress the weed and cause it to expend nutrient stores in its root system. The best management plan includes cultural controls combined with mechanical and/or chemical control techniques.

Scotch Thistle Onopordum Acanthium Canada Thistle Cirsium Arvense

Scotch thistle is a branched, robust biennial (or sometimes annual) that often grows 8 feet or more in height and 6 feet in width. Main stems may be up to 4 inches wide at the base. Stems have vertical rows of prominent, spiny, ribbon-like leaf material or "wings" that extend to the base of the flower heads. Leaves, which are armed with sharp, yellow spines, are up to 2 feet long and 1 foot wide. Upper and lower leaf surfaces are covered with a thick mat of cotton-like or woolly hairs, which give the foliage a gray-green appearance. Plants flower in mid-summer. The globeshaped flower heads are borne in groups of 2 or 3 on branch tips. Flower heads are up to 2 inches in diameter, with long, stiff, needle-like bracts at the base. Flowers range from dark pink to lavender. Seeds are smooth, slender, and plumed.

Canada thistle (Cirsium arvense) is an aggressive, creeping perennial weed that infests crops, pastures, rangeland, roadsides and noncrop areas. Generally, infestations start on disturbed ground, including ditch banks, overgrazed pastures, tilled fields or abandoned sites. Canada thistle reduces forage consumption in pastures and rangeland because cattle typically will not graze near infestations. Canada thistle is a creeping perennial that reproduces from vegetative buds in its root system and from seed. It is difficult to control because its extensive root system allows it to recover from control attempts. Combining control methods is the best form of Canada thistle management. Persistence is imperative so the weed is continually stressed, forcing it to exhaust root nutrient stores and eventually die. See page 53 for more detailed information. Dalmatian toadflax Linaria dalmatica Diffuse knapweed Centaurea diffusa

Toadflax is a menace throughout the northwest and infestations range from light to heavy, with some densities ranging up to 100% of the existing vegetation. This may be explained through the plant's ability to inhibit other plant species from becoming established-this is known as an "El Nino" effect or allelopathic condition. Dalmatian very much resembles its sister species yellow toadflax (Linaria vulgaris); however, the Dalmatian species has wider leaves which clasp the stem.

Dalmatian toadflax, a native of the Mediterranean region, was introduced to the Americas in 1900 as an ornamental. The pioneers loved this plant because of its lavish, bright yellow flowers and its durability. Toadflax was easy to establish and homesteaders liberally landscaped their properties with this drought resistant plant, continually spreading it in their migration to the Western frontier. Also, burial sites were often adorned with toadflax to give everlasting beauty and tranquility. Their legacy has prevailed, proving to be notorious in nature. Diffuse knapweed is a member of the Aster family, Thistle tribe.

Diffuse knapweed was introduced from Europe and is a biennial or short-lived perennial forb which reproduces only by seed. The plant usually produces a single main much-branched stem that is 1-1/2 to 2 feet tall. It is known to have more than one stem produced from one rosette. A basal rosette of leaves is present in young plants with each leaf divided into narrow segments. When leaves are young, a thin nap is present. Stem leaves in the mature plant become much reduced as you ascend to the tip and alternate one per node. Flowers are mostly white, sometimes purple, and are located on each branch tip. The bracts surrounding each flower bear 4 to 5 pairs of lateral spines and one, long terminal spine. Diffuse knapweed can resemble spotted knapweed with the black tipped bracts. The difference is the sharp spine at the end of the bract that is characteristic of diffuse knapweed.

Leafy Spurge Euphorbia Esula Musk Thistle Carduus Nutans

Leafy spurge (Euphorbia esula L.) is a creeping, herbaceous perennial weed of foreign origin that reproduces from seed and vegetative root buds. It can reduce rangeland cattle carrying capacity by 50 to 75 percent. About half of this loss is from decreased grass production. Cattle won't graze in dense leafy spurge stands and these areas are a 100% loss to producers. Leafy spurge is difficult to control. Its extensive root system has vast nutrient stores that let it recover from control attempts.

Combine control methods into a system to achieve best results. See page 55 for more detailed information.

Musk thistle is an aggressive weed of foreign origin that occurs in pastures, rangeland, roadsides and non-crop areas. It is a biennial weed, although occasionally it is an annual. Because musk thistle reproduces solely from seed, the key for successful management is to prevent seed production. Musk thistle is a biennial weed that reproduces only from seed. The key to successful musk thistle control is to prevent seed production. Apply herbicides such as Tordon, Vanquish/Clarity or 2,4-D to musk thistle rosettes in spring or fall. Apply Ally or Telar up to the early flower growth stage. Combine control methods into a management system for best results. Germination and seedling establishment are correlated with moisture and light. Thus, more seeds germinate and establish plants in open pastures and other degraded areas. Vigorously growing grass competes with musk thistle, and fewer thistles occur in pastures where grazing is deferred. However, musk thistle also can become a problem in pasture or rangeland that is in good condition.

Foxtail Setaria Glauca Black Medic Medicago Lupulina

Foxtail is a summer annual grass with wider blades and a lighter green color than bluegrass. It is also faster growing than bluegrass. Seedheads may form despite regular mowing. Foxtail is much less prevalent when turfgrass has good density. Re-sod or reseed bare spots.

Control A pre-emergent herbicide (benefin + trifluralin, dithiopyr, oxadiazon, pendimethalin or prodiamine) applied correctly and at the proper time should provide control. Do not use a preemergent herbicide on a newly-seeded or sodded lawn or when overseeding a lawn. Postemergent herbicides that will kill foxtail seedlings: MSMA, DSMA, MAMA or fenoxaprop. Black medic is an annual, biennial or short-lived perennial. A legume, it is closely related to alfalfa. It is most often found in lawns having low fertility. Often called Japanese clover, this plant has small yellow flowers and a deep taproot. It can be pulled from moist soil without difficulty. Keeping your lawn healthy and dense with proper watering, mowing, and fertilization will discourage invasion by black medic.

Control Triclopyr + 2,4-D combination herbicides or other 2,4-D combination herbicides are more effective than 2,4-D alone. Yellow Alyssum is a member of the mustard family. An exotic imported from Europe, it threatens native grasses. The plants grow from 4 to 8 inches in height and branch at the base. The leaves are narrow, strap-like, and covered with star-shaped hairs. In spring, yellow flowers are borne in racemes. Seeds are formed in silicles that are round, flattened and pointed at the top, much like a bellows. In summer, seeds drop out of the center of the pods, leaving a skeleton that resembles a pair of "spectacles."

Control Yellow alyssum has a taproot that can be easily hand-pulled from moist soil in the garden. Post-emergent treatment with 2-4,D, or triclopyr is effective in controlling this weed in turfgrass. Yellow alyssum can be spot treated with glyphosate (Round-up, Kleen-up). Yellow Alyssum Alyssum alyssoides Shepherd's Purse Capsella bursa-pastoris Shepherd's Purse Another member of the mustard family, Shepherd's purse can produce over 33,000 seeds per plant. The lower leaves are petioled and deeply lobed, quite similar to that of a dandelion. The upper leaves are irregularly toothed and clasping. White flowers appear in clusters at the top of a flowerstalk that can reach 6 to 18 inches in height. The flowers are only about 2mm across. Seeds are contained in silicles that are notched triangles and resemble the purses once carried by shepherds.

Control Shepherd's purse has a slender taproot and can be easily hand-pulled or hoed from moist soil in flower and vegetable gardens while plants are young. In lawns, post-emergent herbicides provide easier control of Shepherd's purse than pre-emergents.

Spot treat by spraying individual plants, rather than applying a weed and feed over the entire lawn. If you would rather not spray, the herbicide can be brushed on instead. Apply the herbicide in mid-spring and again in mid-autumn if a new crop of shepherd's purse emerges. These herbicides are most effective when temperatures are between 60-80°F. Do not spray if temperatures are projected to exceed 85°F within the next 48 hours. Choose a time when no rain is forecast for at least 24, and preferably 48, hours.

To avoid herbicide drift, spray only when the air is still. Drift can harm or kill desirable broadleaf plants such as flowers, vegetables, trees and shrubs. Post-emergent herbicides should be labeled for use in turfgrass and contain a combination of 2,4-D and MCPP or 2,4-D, MCPP and dicamba. Do not apply any product containing dicamba underneath the canopy of young trees, near shrubs or close to gardens as it can be absorbed by their roots.

Blue Mustard Chorispora Tenella Henbit Lamium Amplexicaule

Blue mustard is a winter annual that germinates in the fall and produces a rosette with deeply lobed leaves, similar in appearance to a dandelion. Blue mustard bears purple or blue flowers at the top of the plant in March through April. Leaves on the flowering stems are coarsely toothed and have wavy margins. The plant may grow from 1 to 1 1/2 feet in height. Two-inch long, beanlike seedpods (siliques) that resemble "beaks" mature in early summer.

Control Herbicides are most effective if applied before weeds start to bolt in the spring. In the spring, while it is actively growing, this weed can be controlled with an application of 2,4-D. Henbit is a winter annual occasionally found in lawns in early spring. The lower leaves have a stalk while the upper leaves clasp the stem. Stems are square, like other members of the mint family. All the leaves are coarsely toothed and opposite from each other. Flowers appear in May and are about one-half inch long, trumpet-shaped, pinkish white to purple, and form just above upper leaves. This weed is more often found in buffalograss than in bluegrass. Newly-seeded bluegrass and established bluegrass lawns with poor density may have some henbit. Control Henbit has a taproot and is easily pulled from moist soil. Heavy infestations can be controlled with triclopyr + clopyralid, 2,4-D or 2,4-D combination herbicides; at or prior to flowering. Fall application of a pre-emergent herbicide (dithiopyr, isoxaben, pendimethalin or prodiamine) will prevent henbit germination.

White Clover is a perennial that forms creeping runners. These runners will root at nodes. Many people like clover in lawns, while others find white flowers and the bees they attract objectionable. Clover is a legume which fixes nitrogen, so it is often found in lawns having low fertility. To discourage it, increase turf density with proper watering, mowing and fertilization. Control Triclopyr + 2,4-D combination herbicides or other 2,4-D combination herbicides can be used. 2,4-D alone is not effective.

Buffalo bur, sometimes called Kansas thistle and prickly nightshade, is a taprooted annual weed. It bears long, yellow spines on stems, leaves, and flower heads and can grow up to 2 feet high. Drought resistant, its highest occurrence is in dry, exposed soil. The oblong leaves are 2-3 inches long with deep rounded lobes and are covered with very dense, stiff, and sharp spines. Bright yellow flowers can be seen in summer. In the fall, berries up to 3/8 inch in diameter are enclosed in the dried flower parts and are filled with black, wrinkled, flat pitted seeds. Control of this plant is important as it is a host for the Colorado potato beetle. When mature, the main stem breaks near the ground and the plant rolls like a tumbleweed, widely scattering the 8500 seeds that each plant produces. Buffalo bur can be pulled when the soil is moist. Be sure to wear gloves to avoid injury from the spines.

Control Herbicides should be applied between late bud to early flower. Dicamba, Triclopyr and 2,4-D can be effective in controlling Buffalo bur. Glyphosate in a 2% solution can be applied as a spot treatment.

Puncture vine is a prostrate, mat-forming summer annual. It has small leaflets and small yellow flowers with 5 petals. Fruits containing seeds are a sharp, spiny burr that can easily puncture a bicycle inner tube (or gardener's skin!). This weed is found only in thin, less vigorous turf given insufficient water. It will pull easily out of moist soils. Be sure to wear gloves to protect your hands from the burrs. You can discourage the growth of puncture vine by increasing the turf density. Control A 2,4-D combination herbicide will work best on younger plants. Pre-emergent herbicides such as trifluralin + benefin or dithiopyr applied in spring can provide some control.

Oxalis Oxalis corniculata The leaves of oxalis, also called creeping woodsorrel, have a shamrock appearance and the plant is often mistaken for a clover. At night, or on cloudy days, the leaves may fold up. With the arrival of cooler weather in the fall, leaves turn purplish in color. Occasionally, some plants may have purple leaves all year round. Oxalis is a prostrate, creeping perennial weed with stems that will take root where they touch the ground. Flowers are small and yellow. When mature, fruits explode, scattering seed several feet away. This plant is more common in thin, less vigorous turfgrass that is given too-frequent, light irrigation. It can be discouraged by increasing the density of turfgrass using good cultural practices.

Oxalis Control 2,4-D combination herbicides applied in spring and/or fall give marginal levels of control. Triclopyr + clopyralid control is fair to good. Pre-emergent herbicides (pendimethalin, dithiopyr, isoxaben, prodiamine) applied 2-3 successive years in late March-early April can provide some control. White clover Trifolium repens Buffalo Bur Solanum rostratum Puncture vine Tribulus terrestris Oxalis Oxalis corniculata Redroot Pigweed Amaranthus retroflexus Dandelion Taraxacum officinale

Redroot pigweed is an annual weed commonly found in waste areas and disturbed soils. It can attain a 2-4 foot height. The lower stems are reddish in color. Flowers are small and green, in bristly terminal and axillary clusters. A prolific seed producer, pigweed will produce up to 100,000 seeds per plant. Seedlings are easily pulled or hoed from the garden. Older plants are more difficult to pull out and may scatter seeds in the process. Mulch will prevent seeds already in the soil from germinating.

Control When found in the lawn, pigweed can be spot-treated with herbicides containing MCPP, MCPA, and dicamba. In gardens, glyphosate (Round-up, Kleen-up) can be used, taking care not to get any of the chemical on desired plants. Pre-emergents containing oryzalin, pendimethalin, or trifluralin can be effective. When using a herbicide, be sure to read the label carefully and follow directions closely.

Dandelion The Dandelion is a perennial with an extensive taproot. Its yellow flowers can develop anytime between March and November and are followed by fluffy seedheads. More prevalent under low turf density, dandelion growth can be inhibited by increasing the turf density. Dandelions can be dug out with special tools, but any part of the root that is left is capable of regenerating a plant.

Control A 2,4-D or 2,4-D combination herbicide is most-effective and should be used in spring and fall. Always read the label before applying any pesticide.

Curly dock is a perennial weed in the buckwheat family. Fairly pleasant tasting, the leaves are very rich in vitamins, especially vitamins A and C, and can be eaten raw or cooked. The roasted seed has been used as a coffee substitute. It is also a very important food plant for the caterpillars of many butterflies. In the spring, basal leaves emerge from a stout taproot. These elongated leaves have wavy margins, thus the name "curly" dock. In summer, the plant has reddish, rigid stems, 2-4 feet tall. Flower stems have greenish flowers.

This is a tenacious perennial weed that is found in lawns throughout the United States. Its large taproot grows deep into the soil, which enables it to thrive in times when grass may be suffering from heat and lack of moisture. In fall, winged fruits form on the flowering stems that are reddishbrown in color. Curly dock is not easily pulled because of the deep taproot, and portions of this root left in the ground will regenerate.

Control Combination herbicides containing mecoprop, dicamba and 2,4-D are effective in the control of Curly dock.

Quackgrass, a creeping perennial, is a very aggressive grass and is considered a noxious weed in most states. It spreads by seeds and invasive rhizomes (underground stems). Its rhizomes are yellow-white, with brown sections, and the ends are sharply pointed.

Control A few quackgrass plants can be spot-sprayed with glyphosate, or individual blades can be painted with glyphosate. Note that glyphosate will kill any bluegrass it contacts. Repeat applications to quackgrass will likely be needed. Renovate severely infested lawn areas. Spray with glyphosate; repeat applications will likely be needed. When the quackgrass has been killed, the areas may be resodded or reseeded. Always read the label before applying any pesticide. Broadleaf Plantain is a low growing perennial. It has broad leaves with prominent veins. The leaves are arranged in a rosette and may smother lawn grass. The flowering spikes normally grow taller than the foliage but may develop below mowing height. Vigorous, thick turfgrass is less susceptible to invasion.

Control Triclopyr + 2,4-D or 2,4-D alone or 2,4-D combination herbicides should control plantain. Always read the label before applying any pesticide.

Quackgrass Elytrigia repens Curly Dock Rumex crispus

Broadleaf Plantain Plantago major Common Lambsquarters Chenopodium album

Purslane Portulaca oleracea Crabgrass Digitaria sanguinalis

Common Lambsquarters is a broadleaf summer-annual weed that can be found anyplace the soil has been disturbed. The growth habits of the common lambsquarters vary with its location. If growing along the road or in an open field, it may reach three or four feet in height. Yet those plants found in lawns may produce seeds when only a couple of inches tall. Lambsquarters prefer rich soil and are often found in vegetable gardens. The first two true leaves that appear are opposite and ovate with smooth edges. Later leaves are alternate, arrowhead-shaped and with unevenly-toothed edges. At the top of the plant, the leaves become long and narrow. All of the leaves are pale green and covered with white, mealy granules that give the appearance of frost. Stems are erect and often have light-green or red lines.

Flowers are small, inconspicuous, yellowgreen, and occur in clusters at the tips of branches and upper leaf axils. These flowers occur from July to September. Black seeds are contained in an utricle by a thin papery covering. Control The best methods of weed control in the home vegetable garden are mulching, handpulling, rototilling, hoeing and preventing the weeds from going to seed.

Because of its short, branched taproot, lambsquarters can be easily hand-pulled from moist soil. Prevention by use of good cultural habits should be the first line of defense in eliminating broadleaf weeds such as lambsquarters from lawns. Pre-emergent herbicides such as trifluralin (Preen) can be used to prevent germination of weed seeds. Post-emergent herbicides effective against broadleaf weeds are 2,4-D, MCPP and dicamba (sold under many brand names) and combination formulas

(Trimec). Read labels to be sure the herbicide is effective against the weeds you want to eliminate and follow the directions carefully.

Purslane is a summer annual, found only in newly seeded or thinning, non-vigorous lawns and sidewalk cracks. It has thick, succulent, rubbery leaves and small yellow flowers. The plant is sprawling and prostrate. It is easily pulled when the soil is moist.

Control Post-emergent herbicide use is more effective when plants are young as purslane is difficult to kill with an herbicide when larger. Always read the label before applying any pesticide. Crabgrass is a summer annual grass with wider blades and a lighter green color than bluegrass. It is low growing, prostrate, and often has reddish-purple stems. It forms seedheads below mowing height. The seedheads are composed of slender, fingerlike spikes. Crabgrass is less prevalent when turf has good density. In particular, mowing too low promotes crabgrass seed germination. Maintain mowing heights of 2.5-3 inches.

Control A pre-emergent herbicide (benefin + trifluralin, dithiopyr, DCPA, oxadiazon, pendimethalin, or prodiamine) applied correctly and at the proper time should provide control. Do not use a pre-emergent herbicide on a newly seeded or sodded lawn or when overseeding a lawn. Fenoxaprop and other post-emergent "crabgrass killer" (MSMA, DSMA, MAMA) sprays are not effective unless crabgrass plants are immature, young seedlings. Always read the label before applying any pesticide.

Poison Hemlock (Conium maculatum), also known as poison parsley or spotted parsley, is an erect biennial weed that can grow 6 to 10 feet tall. Originally imported from Europe as an ornamental plant, its spread across North America has been rampant. All parts of this plant are poisonous. The leaves and flowers of Poison Hemlock are similar to those of parsnips and carrots. In fact, many deaths have occurred as a result of people mistaking it for the edible species of the carrot family. The use of hemlock as a poison goes back many years in history. It is said that the ancient Greeks used it to poison their enemies and political prisoners. Socrates, condemned to die as a political prisoner in 329 B.C., drank the juice of the hemlock plant and committed suicide. Native Americans were known to dip their arrows in hemlock. Poison Hemlock is an erect plant with smooth, hollow stems that are covered with purple spots. Its shiny green leaves are pinnately compound, multi-stemmed and fern-like in appearance. Flowers are showy, white umbrella-like clusters that occur during June to July. The fruit is small, flat, grayish-green, and matures in August to September.

Control The large, fleshy white taproot can be easily pulled from moist ground when plants are small. Several herbicides are effective in controlling Poison Hemlock. Glyphosate (Roundup) can be used on newly emerged sprouts. Other post-emergents that are most effective in early spring are 2,4-D, 2,4-DB and MCPA. 2,4-D should be combined with a wetting agent when applied.

Because of the large number of seeds that may have been produced, repeated herbicide applications may be necessary. The frequency of application varies with the herbicide. Read labels carefully and follow directions as given. Biological control has been effective using the Hemlock moth (Agonopterix alstroemericana). The larvae of this moth feed on the leaves, young stem tissue, flowers and seeds of Hemlock plants, causing severe defoliation and death of the plant.

The plant typically grows in open areas, such as roadsides and fields, at elevations between 1000-5500 feet. The Silverleaf Nightshade is poisonous; however, Native Americans have used the crushed berries of the plant to curdle milk since they contain a protein-digesting enzyme. Silverleaf Nightshade typically blooms between May-October and produces a bluish purple, star shaped flower with yellow anthers in the middle to 1.5 inches in width. The flower is followed by a 1/2 inch diameter yellow berry.

Control Cobra may be used in combination with MSMA, Bladex and Karmex to aid in control of certain weeds. Consult specific product labels for recommendations and precautions. Bamboo-Phyllostachys: From the Greek "phyllon", leaf, and "stachys", a spike, referring to the leafy bloom spike (inflorescence). Subfamily: Bambusoideae A diverse genus of about 15 species of evergreen running bamboo primarily native to temperate and subtropical China and Japan, Phyllostachys includes several popular bamboo of gardens as well as varieties valued for timber and for edible shoots. The rounded culms display distinctive grooves or compressed areas on the branching sides, making this genus fairly easy to recognize and distinguish from other bamboo. Their dense evergreen foliage, attractively marked and colored culms, and fast growth have made these running bamboos favorites for creating large groves, for screening and hedging, and for stabilizing rough slopes or streamsides. On small properties the spreading roots of Phyllostachys may be confined as desired by installing appropriate barriers at planting. Culture: Phyllostachys species thrive in sun or light shade if provided with abundant moisture and rich soil. Where drought may be expected or in hot interior climates, some shading would be beneficial for most varieties. Established plants withstand flooding.

Control Bamboo can be controlled in a small area (a normal city yard) simply by mowing or stepping on the young shoots. New shoots could also be used for culinary purposes and cooked, much like asparagus. Keep in mind that running bamboo only puts up new shoots for a short time each year, and when they are "shooting", the new plants are very fragile and easy to destroy. To really slow the spread of the rhizomes, the grove can be root pruned at any time. (Best time seems to be in the fall or early winter). This is done with a sharp, narrow spade (8" wide, flat spade works quite well) by "trimming" the rhizomes on all sides which need to be controlled. Push the shovel straight down to its maximum depth so as to cut through any rhizomes encountered, (8-10 inches is usually sufficient). Keep moving the spade one width at a time to the right or left, cutting a line until all the rhizomes have been cut. Yes, you may miss a few, but you can get them next year. You do not have to dig up the whole root system and rhizomes. Once you have severed it from the mother plant and cut off any new shoots, you will have deprived it of all food sources and it will die off. Glyphosate (Roundup, Kleenup) Systemic type: Controls a wide range of annual and perennial grasses and broadleaf plants, but in many cases requires very precise timing to be effective.

Common mallow is most frequently found in newly seeded lawns or lawns that are stressed and lack density. It can be an annual or biennial. Mallow has a deep taproot but can be easily pulled from moist soil. The foliage resembles that of the geranium. The flowers of common mallow are pinkish-white and the fruits look like small, round cheeses.

Control Increasing turf density with proper mowing, fertilization, watering and other cultural practices can help in the control of this weed. Post-emergent herbicides are only marginally effective. Triclopyr + clopyralid or triclopyr alone are suggested. Common groundsel is an early season weed. A prolific seed producer. Seeds are produced within several weeks of germination, and there are several generations within the same year. This weed likes moist soil and is often found in well-irrigated areas such as lawns and flower beds. Control A dense, healthy turf will prevent seeds from taking root in the lawn. Turf density can be increased with proper mowing, fertilization, watering and other cultural practices. Good drainage will also help to discourage the growth of common groundsel. The plants can be easily pulled by hand from moist soil. Be sure to pull and dispose of them before they set seed, as seed can mature in opened flowers even after the plants have been killed. If there is heavy infestation, spot treat with a post-emergent herbicide containing glyphosate (Roundup, Kleenup).

Russian Thistle Salsola Iberica Milkweed Asclepias Syriaca

Redstem Filaree is a prostrate broadleaf winter annual or biennial weed. A member of the geranium family, it has fern-like, or feathery appearing foliage in a rosette. The stems are reddish and there is a large, white taproot.

Control Appearing more frequently in newly seeded lawns or thin, stressed lawns, this minor turf weed can be eliminated by increasing the density of the lawn. Newly seeded lawns will develop more density with time and will then crowd out this weed. It is often seen in buffalograss lawns. The entire plant can be easily pulled out of moist ground. Post emergent herbicides such as 2,4-D or 2.4-D combinations are only marginally effective in controlling Redstem filaree. London rocket is a European native weed belonging to the mustard family, and is one of the first winter weeds to appear in southern Arizona. It is abundant in irrigated land in crops such as alfalfa and small grains, in gardens, citrus orchards, pastures, and along roadsides. London rocket is a bright green fleshy winter annual. The stems branch from the base 1 to 3 feet high. It has a coarse taproot. Small, yellow flowers are borne on slender stalks in small clusters at the stem tip.

Russian thistle was introduced from Russia and is an annual which reproduces by seed. It is a round, bushy, much-branched plant growing 1 to 3-1/2 feet high. The branches are slender, succulent when young, and woody when mature. Russian thistle grows on dry plains, in cultivated fields, roadsides, and waste places, chiefly in grain-growing areas. At maturity, the plant breaks off at the base and because of its round shape, becomes an excellent tumbleweed. Russian thistle is an annual. A single plant may produce 20,000 to 50,000 seeds. Seeds germinate quickly, even after brief or limited precipitation.

The plants are well branched, round bushy that grow 1/2 to 3 feet high. The stems have reddish or purple strips. The leaves are alternate, long, and very thin or needle-like. Flowers are small and inconspicuous and develop in the upper leaf axils. Each flower has a pair of spiny bracts. Mature plants break off at ground level and tumble, spreading the seeds. This species is a host of sugar beet leafhopper, which transmits curly top of sugar beets. This weed is common in drier areas and in dryland cultivation systems. It is found in overgrazed rangeland, disturbed waste areas and in wheat, sugar beet, potato, and mint crops east of the Cascade Mountains.

Control Control of Russian thistle is difficult. There have been numerous attempts through the years to import biological control agents, but none have been successful.

Pre-emergent herbicides are applied to the soil before the weed seed germinates and are usually incorporated into the soil with irrigation or rainfall. The most effective pre-emergent herbicides are Aatrex (atrazine), Velpar (hexazinone), Devrinol (napropamide), Telar (chlorsulfuron), Oust (sulfometuron), Princep (simazine) and Hyvar (bromacil). Other preemergent herbicides that are registered but only moderately effective in controlling Russian thistle are Surflan (oryzalin), Treflan (trifluralin), Prowl (pendimethalin), Endurance (prodiamine), Lasso (alachlor), Predict (norflurazon), and Kerb (pronamide).

Herbicide-resistant biotypes of Russian thistle have evolved in only a couple of years following treatment with Telar (chlorsulfuron) or Oust (sulfometuron). Avoid repeated use of a single herbicide or of herbicides that have the same mode of action to prevent the evolution of herbicide-resistant populations.

Post-emergent herbicides are applied to plants, but timing is critical. For best results, these herbicides must be applied while the weed is in its early growth stages, preferably the early seedling stage, before it becomes hardened and starts producing its spiney branches. Do not use post-emergent herbicides to try to control the mature seed (either on the plant or on the ground) as they are not effective for this purpose. Also, the later spiney stage of Russian thistle is not readily controlled by any post-emergent herbicide. If rain or irrigation occurs after a postemergent application, additional seedlings may emerge and require future treatments. Postemergent herbicides that are effective when properly applied include Banvel or Vanquish (dicamba), Roundup (glyphosate), 2,4-D and Gramoxone (paraquat).

Milkweed plants, members of the Asclepias family, are the only host plant for the monarch and queen butterflies. The adult females seek out these plants on which they lay their eggs. The caterpillars that hatch will remain on the plants and eat the leaves until they enter the pupal stage, then emerge as adult butterflies. It is a perennial herb with long-spreading rhizomes. Stems are stout, erect, to 2 m tall, with short downy hairs and milky juice; leaves opposite, oblong, rounded, 1-2.6 dm long, 0.4-1.8 dm broad, with prominent veins; upper surface smooth, lower covered with short white hairs and strong transverse nerves. Flowers sweet-smelling, pink to white, in large, many-flowered, axillary and apical bell-like clusters; corolla lobes 6-9 mm long, hoods 3-4 mm high; follicle grayish, hairy, with soft spiny projections, 1-3 mm high, slenderly ovoid, 2.5-3.5 cm thick. Seed brown, flat, oval, 6 mm long, 5 mm wide, with a tuft of silky white hairs apically (Reed, 1970)

Control Non-Cropland--Established stands of common milkweed can be controlled with herbicides. On non-cropland (i.e., roadsides, railroad rights-of-way, etc., but not idle land or grazing land), Amitrol-T, Roundup or Tordon would provide control. Herbicide coverage of the common milkweed is incomplete with this method, resulting in reduced effectiveness compared to a sprayer. Used over several years, however, common milkweed populations can be reduced with Roundup applied in a wiper applicator. Controls: Amino Trizole;Amitrol-T, Roundup, 2,4-D+Banvel,and Tordon

INVASIVE PLANT SPECIES

Invasive species can be plants, animals, and other organisms (e.g., microbes). Human actions are the primary means of invasive species introductions.

As Defined by the Oregon Invasive Species Council

Next to habitat lost to land development and transformation, invasive species pose the greatest threat to the survival of native biota in the United States, and many other areas of the world. Invasive species are those plants, animals, and microbes not native to a region which, when introduced either accidentally or intentionally, out-compete native species for available resources, reproduce prolifically, and dominate regions and ecosystems. Because they often arrive in new areas unaccompanied by their native predators, invasive species can be difficult to control. Left unchecked, many invasives have the potential to transform entire ecosystems, as native species and those that depend on them for food, shelter, and habitat disappear.

ALIEN PLANTS

Alien plants, also known as exotic, non-native, or non-indigenous plants, are species intentionally or accidentally introduced by human activity into a region in which they did not evolve. Many alien species are well known and economically important in agriculture and horticulture, such as wheat, soybeans, and tulips. Alien species, whether plant or animal, often do not become established outside of cultivation and, if they do, they usually have few impacts on natural communities. Invasive alien plants, however, escape cultivation and become agricultural pests, infest lawns as weeds, displace native plant species, reduce wildlife habitat, and alter ecosystem processes. Across the country and around the world, invasive alien plants and animals have become one of the most serious threats to native species, natural communities, and ecosystem processes.

They also exact a costly toll from-human economies that depend on resources and services provided by healthy ecosystems. Examples include destruction of vast areas of western rangelands, clogging of important waterways, and increased costs in maintaining open powerline rights-ofway. Of the 4,000 alien plant species introduced to the United States that have escaped cultivation, approximately 400 are serious invaders. Half this total were introduced for horticultural uses.

Others arrived accidentally in seed mixes, packaging materials, ships ballast, and by other means. Invasive plants now infest more than 100 million acres. One study estimated that from 1901 to 1991, economic losses in the U.S. caused by 15 invasive plant species (not including agricultural weeds) were $603 million (Office of Technology Assessment, 1993).

Once thought to be a problem only on farms or in lawns, invasive plants are now recognized as a threat to natural areas, parks, forests, and other sites in a more or less natural state. Land managers, weed scientists, foresters, ecologists, and other conservationists are joining together to face this challenge in ways that help conserve native species and natural communities and protect environmental quality.

Invasive alien plants typically exhibit the following characteristics:

- o Rapid growth and maturity
- o Prolific seed production
- o Highly successful seed dispersal, germination and colonization
- o Rampant vegetative spread
- o Ability to out-compete native species
- o High cost to remove or control

An invasive plant infestation is like a slow motion explosion, which, if left unchecked, may severely alter a site's natural, economic, aesthetic, and other cultural values. Management of invasive species while maintaining these values can appear to be a complicated and unending task.

For this reason, planning and prioritizing are crucial. By articulating clear goals, gathering the best available information, and prioritizing actions based on the significance of an infestation's impacts and feasibility of control, land managers can identify how their time, effort, and money can most effectively be applied. Invasive species present a difficult challenge with no quick and easy solutions. Many unknowns exist regarding control methods and their efficacy, in addition to limited budgets for managing invasive plants. Sometimes, the best course of action may be to do nothing.

COMMON ALIEN PLANTS WHICH ARE USED AS HERBS

Goutweed (Aegopodium podagraria)

Garlic mustard (Alliaria petiolata)

Giant reed (Arundo donax)

Musk thistle (Carduus nutans)

Asiatic sand sedge (Carex kobomugi)

Spotted knapweed (Centaurea biebersteinii)

Yellow starthistle (Centaurea solstitialis)

Canada thistle (Cirsium arvense)

Leafy spurge (Euphorbia esula)

Cogon grass (Imperata cylindrica)

Chinese lespedeza (Lespedeza cuneata)

Purple loosestrife (Lythrum salicaria)

Japanese stilt grass (Microstegium vimineum)

Eurasian watermilfoil (Myriophyllum spicatum)

Burma reed (Neyraudia reynaudiana)

Fountain grass (Pennisetum setaceum)

Japanese knotweed (Polygonum cuspidatum)

Lesser celandine (Ranunculus ficaria)

Annual bastard-cabbage (Rapistrum rugosum)

Common mullein (Verbascum thapsus)

Annual bastard-cabbage

Goutweed Aegopodium podagraria L.

Native Range: Most of Europe and northern Asia, to eastern Siberia

Deseription: Goutweed, also known as bishop's-weed and snow-on-the-mountain, is an herbaceous perennial plant in the carrot family (Apiaceae, formerly the Umbelliferae). It is one of several species of Aegopodium, native to Europe and Asia. Most leaves are basal, with the leafstalk attached to an underground stem, or rhizome.

The leaves are divided into three groups of three leaflets, making it "triternate." The leaflets are toothed and sometimes irregularly lobed. Foliage of the "wild" type is medium green in color; a commonly planted variegated form has bluish-green leaves with creamy white edges. Sometimes reversion back to solid green or a mixture of solid green and the lighter variegated pattern occurs within a patch. Small, white, five-petaled flowers are produced in mid-summer. Flowers are arranged in flat-topped clusters (called compound umbels) and are held above the ground on a leafy stem up to about 3 feet tall. The seeds are small and elongate, similar in size and shape to carrot seeds, and ripen in late summer. In contrast to the dense foliage cover produced by goutweed, flowering shoots are uncommon in densely shaded areas.

The rhizomes of goutweed are long, white, and branching, superficially resembling those of quackgrass (Elytrigia repens, also known as Agropyron repens). Patches of goutweed typically form a dense canopy and can exclude most other herbaceous vegetation. Because of this, it is often used as a low-maintenance ground cover.

Ecological Threat: Goutweed is an aggressive invasive plant that forms dense patches, displaces native species, and greatly reduces species diversity in the ground layer. Goutweed patches inhibit the establishment of conifers and other native tree species as well.

Distribution in the United States: Goutweed is currently known to occur in twentynine states in the mid-Atlantic, Northeast and Northwest and is reported to be invasive in natural areas in Connecticut, Michigan, New Jersey, Pennsylvania, Vermont, and Wisconsin.

Habitat in the United States: Goutweed is an ecologically versatile species. It is found in old gardens and flowerbeds, around shrubs and other plantings, and in a variety of other disturbed habitats such as felled forests, abandoned fields, and pastures. In Eurasia, goutweed is primarily a species of deciduous and southern boreal forests, and it expands aggressively in similar habitats in North America. Goutweed appears to do best on moist soil and in light to moderate shade, but is highly shade-tolerant and capable of invading closed-canopy forests.

Background: Goutweed was apparently first brought to North America as an ornamental during the early stages of European settlement and was well established in the U.S. by 1863. In parts of Russia, the leaves are sometimes used as a salad ingredient and potherb in the spring.

Biology & Spread: Goutweed is vigorous, rhizomatous perennial that spreads mainly by vegetative means. Patches increase in size through extension of the rhizome system. The flowers are pollinated by a wide variety of insects, including beetles, bees, and especially small flies. The seeds have no apparent morphological features that would facilitate dispersal. Goutweed seeds require cold stratification to germinate. While established goutweed plants are highly competitive in shaded environments, seedlings generally need recently disturbed soil and rather bright light in order to survive. Goutweed apparently does not form a long-lived seed bank, and the seeds generally germinate the year after ripening. Establishment of goutweed seedlings in the shade is rare. The primary vector for dispersal to new areas is humans. Most goutweed colonies spread to neighboring natural areas from intentional plantings, or by the dumping of yard waste that includes discarded rhizomes.

Current Management Approaches: A variety of methods are available for controlling goutweed, depending on the extent of the infestation and the amount of time and labor available. Regardless of the control method used, the patch should be carefully monitored periodically for a few years. New shoots should be dug up and destroyed. Once goutweed control has been achieved, revegetation with native or non-invasive exotic plant materials is recommended. This is particularly important on sites where erosion is a concern or where other invasive species are likely to colonize the site if left alone.

Biological: There are no biological control organisms currently available for Aegopodium podograria in North America.

Chemical: Systemic herbicides such as glyphosate (Roundup) that are translocated to the roots and kill the entire plant are most effective for goutweed control. However, glyphosate is nonspecific and can damage or kill desirable native plants that are accidentally sprayed in the course of treating the goutweed. Contact herbicides are usually ineffective because goutweed readily leafs out again after defoliation.

Manual: Small patches of goutweed can be eliminated by careful and persistent hand-pulling or digging up of entire plants along with underground stems (rhizomes). Pulled plants can be piled up and allowed to dry for a few days before bagging and disposing of them. Be careful to pick up all rhizomes which, if left behind, can re-root and sprout new plants. For large patches, a team of volunteers or use of herbicide is recommended.

Mechanical: Where appropriate, frequent short mowing may control or slow the spread of goutweed in lawns, along roadsides, and other areas.

Physical: Preventing goutweed from photosynthesizing in early spring (at the time of leaf-out) can control the plant by depleting its carbohydrate reserves. This can be accomplished by covering the patch with black plastic sheeting when the leaves start to emerge from the ground in the spring, and leaving it in place through the summer. A more effective option is to cut all plants once they've fully leafed out, using a mower, scythe, or weed-whacker type machine, and then cover the area with plastic. Covering the plants in mid-or late summer, after they have regained substantial starch reserves, is probably much less effective.

GARLIC MUSTARD ALLIARIA PETIOLATA [BIEB] CAVARA & GRANDE

Native Range: Europe.

Description: Garlic mustard is a cool season biennial herb in the mustard family (Brassicaceae) with stalked, triangular to heart-shaped, coarsely toothed leaves that give off an odor of garlic when crushed. First-year plants appear as a rosette of green leaves close to the ground. Rosettes remain green through the winter and develop into mature flowering plants the following spring. Flowering plants of garlic mustard reach from 2 to 3-1/2 feet in height and produce buttonlike clusters of small white flowers, each with four petals in the shape of a cross.

Recognition of garlic mustard is critical. Several whiteflowered native plants, including toothworts (Dentaria), sweet cicely (Osmorhiza claytonii), and early saxifrage (Saxifraga virginica), occur alongside garlic mustard and may be mistaken for it.

Beginning in May (in the mid-Atlantic Coast Plain region), seeds are produced in erect, slender pods and become shiny black when mature. By late June, when most garlic mustard plants have died, they can be recognized only by the erect stalks of dry, pale brown seedpods that remain, and may hold viable seed, through the summer.

Ecological Threat: Garlic mustard poses a severe threat to native plants and animals in forest communities in much of the eastern and midwestern U.S. Many native wildflowers that complete their life cycles in the springtime (e.g., spring beauty, wild ginger, bloodroot, Dutchman's breeches, hepatica, toothworts, and trilliums) occur in the same habitat as garlic mustard. Once introduced to an area, garlic mustard outcompetes native plants by aggressively monopolizing light, moisture, nutrients, soil and space. Wildlife species that depend on these early plants for their foliage, pollen, nectar, fruits, seeds and roots, are deprived of these essential food sources when garlic mustard replaces them. Humans are also deprived of the vibrant display of beautiful spring wildflowers.

Garlic mustard also poses a threat to one of our rare native insects, the West Virginia white butterfly (Pieris virginiensis). Several species of spring wildflowers known as "toothworts" (Dentaria), also in the mustard family, are the primary food source for the caterpillar stage of this butterfly. Invasions of garlic mustard are causing local extirpations of the toothworts, and chemicals in garlic mustard appear to be toxic to the eggs of the butterfly, as evidenced by their failure to hatch when laid on garlic mustard plants.

- **Distribution in the United States:** Garlic mustard ranges from eastern Canada, south to Virginia and as far west as Kansas and Nebraska.

Habitat in the United States: Garlic mustard frequently occurs in moist, shaded soil of river floodplains, forests, roadsides, edges of woods and trails edges and forest openings. Disturbed areas are most susceptible to rapid invasion and dominance. Though invasive under a wide range of light and soil conditions, garlic mustard is associated with calcareous soils and does not tolerate high acidity. Growing season inundation may limit invasion of garlic mustard to some extent.

Background: Garlic mustard was first recorded in the United States about 1868, from Long Island, New York. It was likely introduced by settlers for food or medicinal purposes.

Biology & Spread: After spending the first half of its two-year life cycle as a rosette of leaves, garlic mustard plants develop rapidly the following spring into mature plants that flower, produce seed and die by late June. In the mid-Atlantic Coastal Plain region, seeds are produced in erect, slender, four-sided pods, called siliques, beginning in May. Siliques become tan and papery as they mature and contain shiny black seeds in a row. By late June, most of the leaves have faded away and garlic mustard plants can be recognized only by the dead and dying stalks of dry, pale brown seedpods that may remain and hold viable seed throughout the summer. A single plant can produce thousands of seeds, which scatter as much as several meters from the parent plant. Depending upon conditions, garlic mustard flowers either self-fertilize or are cross-pollinated by a variety of insects. Self-fertilized seed is genetically identical to the parent plant, enhancing its ability to colonize an area. Although water may transport seeds of garlic mustard, they do not float well and are probably not carried far by wind. Long distance dispersal is most likely aided by human activities and wildlife. Additionally, because white-tailed deer prefer native plants to garlic mustard, large deer populations may help to expand it by removing competing native plants and exposing the soil and seedbed through trampling.

Current Management Approaches: Because the seeds of garlic can remain viable in the soil for five years or more, effective management requires a long term commitment. The goal is to prevent seed production until the stored seed is exhausted. Hand removal of plants is possible for light infestations and when desirable native species co-occur. Care must be taken to remove the plant with its entire root system because new plants can sprout from root fragments. This is best achieved when the soil is moist, by grasping low and firmly on the plant and tugging gently until the main root loosens from the soil and the entire plant pulls out. Pulled plants should be removed from site if at all possible, especially if flowers are present.

For larger infestations of garlic mustard, or when hand-pulling is not practical, flowering stems can be cut at ground level or within several inches of the ground, to prevent seed production. If stems are cut too high, the plant may produce additional flowers at leaf axils. Once seedpods are present, but before the seeds have matured or scattered, the stalks can be clipped, bagged and removed from the site to help prevent continued buildup of seed stores. This can be done through much of the summer. For very heavy infestations, where the risk to desirable plant species is minimal, application of the systemic herbicide glyphosate (e.g., Roundup) is also effective. Herbicide may be applied at any time of year, including winter (to kill overwintering rosettes), as long as the temperature is above 50°F and rain is not expected for about 8 hours. Extreme care must be taken not to get glyphosate on desirable plants as the product is non-selective and will kill almost any plant it contacts. Spray shields may be used to better direct herbicide and limit non-intentional drift.

Fire has been used to control garlic mustard in some large natural settings but, because burning opens the understory, it can encourage germination of stored seeds and promote growth of emerging garlic mustard seedlings. For this reason, burns must be conducted for three to five consecutive years. Regardless of the control method employed, annual monitoring is necessary for a period of at least five years to ensure that seed stores of garlic mustard have been exhausted.

Researchers are investigating potential biological control agents for garlic mustard which may greatly improve the control of this insidious weed.

GIANT REED ARUNDO DONAX L.

Native Range: India

Description: Giant reed, also known as wild cane, is a tall, perennial grass that can grow to over 20 feet in height. Its fleshy, creeping rootstocks form compact masses from which tough, fibrous roots emerge that penetrate deeply into the soil. Leaves are elongate, 1-2 inches wide and a foot long. The flowers are borne in 2-foot long, dense, plume-like panicles during August and September.

Ecological Threat: Giant reed chokes riversides and stream channels, crowds out native plants, interferes with flood control, increases fire potential, and reduces habitat for wildlife, including the Least Bell's vireo, a federally endangered bird. The long, fibrous, interconnecting root mats of giant reed form a framework for debris dams behind bridges, culverts, and other structures that lead to damage. It ignites easily and can create intense fires. Giant reed can float miles downstream where root and stem fragments may take root and initiate new infestations. Due to its rapid growth rate and vegetative reproduction, it is able to quickly invade new areas and form pure stands at the expense of other species. Once established, giant reed has the ability to outcompete and completely suppress native vegetation.

Distribution in the United States: Giant reed is distributed from Arkansas and Texas to California, where it is found throughout the state, and in the east, from Virginia to Kentucky and Missouri and generally southward.

Habitat in the United States: Giant reed becomes established in moist places such as ditches, streams, and riverbanks, growing best in well drained soils where abundant moisture is available. It tolerates a wide variety of conditions, including high salinity, and can flourish in many soil types from heavy clays to loose sands.

Background: Giant reed was probably first introduced into the United States in Los Angeles, California in the early

1800's. Since then, it has become widely dispersed into all of the subtropical and warm temperate areas of the world, mostly through intentional human introductions. Today, giant reed is widely planted throughout the warmer areas of the United States as an ornamental and in the Southwest, where it is used along ditches for erosion control. Giant reed has a variety of uses ranging from music to medicine. Primitive pipe organs were made from it and the reeds for woodwind instruments are still made from its culms, for which no satisfactory substitutes are known. It is also used in basketry, for fishing rods, livestock fodder, medicine, and soil erosion control.

Biology & Spread: Reproduction of giant reed is primarily vegetative, through rhizomes which root and sprout readily. Little is known about the importance of sexual reproduction in giant reed, or about its seed viability, dormancy, and germination, and seedling establishment. Research on these topics may yield some additional improvements in the management of giant reed.

Current Management Approaches: Areas infested with giant reed are best restored through chemical means. Mechanical control (e.g., repeated mowing) may be somewhat effective, but if small fragments of root are left in the soil, they may lead to reestablishment. Systemic herbicides, such as glyphosate (e.g., Rodeo), may be applied to clumps of giant reed, after flowering, either as a cut stump treatment or as a foliar spray. When applying herbicides in or around water or wetlands, be sure to use products labeled for that purpose to avoid harm to aquatic organisms.

Prescribed burning, either alone or combined with herbicide applications, may be effective if conducted after flowering. Once giant reed has been reduced sufficiently, native plants may be seeded or transplanted at the treated site.

MUSK THISTLE CARDUUS NUTANS L.

Native Range: Europe and Asia.

Description: Musk, or nodding thistle, a member of the aster family (Asteraceae), is an aggressive, biennial herb with showy red-purple flowers and painful spiny stems and leaves. Mature plants range in height from 1½ to 6 feet tall, and have multi-branched stems. Leaves are dark green, coarsely lobed, with a smooth waxy surface and a yellowish to white spine at the tip. The large disk-shaped flower heads, containing hundreds of tiny individual flowers, are 1½ to 3½ inches in length and occur at the tips of stems. Flower heads will droop to a 90-degree angle from the stem when mature, hence its alternate name, nodding thistle. Each plant may produce thousands of straw-colored seeds adorned with plume-like bristles.

Ecological Threat: Because musk thistle is unpalatable to wildlife and livestock, selective grazing leads to severe degradation of native meadows and grasslands as wildlife focus their foraging on native plants, giving musk thistle a competitive advantage. Although musk thistle is infrequently found in dense forests, it can colonize areas subjected to natural disturbances such as landslides or frequent flooding. Meadows, prairies, grassy balds, and other open areas are susceptible to invasion.

Distribution in the United States: Musk thistle is found throughout the U.S. except for Maine, Vermont, Florida, Alaska and Hawaii.

Habitat in the United States: Musk thistle grows from sea level to about 8,000 ft elevation, in neutral to acidic soils. It invades open natural areas such as meadows, prairies, and grassy balds. It spreads rapidly in areas subjected to frequent natural disturbance events such as landslides and flooding but does not grow well in excessively wet, dry or shady conditions.

Background: A native of western Europe, musk thistle was introduced into the eastern United States in the early 1800s and has a long history as a rangeland pest in the U.S. It was first discovered in Davidson County, Tennessee in 1942 and has been declared a noxious weed in many states, including North Carolina.

Biology & Spread: Musk thistle is usually a biennial, requiring 2 years to complete a reproductive cycle, but may germinate and flower in a single year in warmer climates. Seedlings emerge in mid to late July and develop into a rosette where plants can reach 4 feet in diameter. Plants overwinter in the rosette stage until they begin to bolt in mid-March. During the bolting stage plants form multi-branched stems to a height of 6 feet. The number of seedheads per plant is site-dependent and ranges from about 24 to 56 on favorable sites and 1 to 18 on less favorable sites. Flowers emerge in early May to August and seed dissemination occurs approximately one month after the flowers form. A single flower head may produce 1,200 seeds and a single plant up to 120,000 seeds, which may be windblown for miles. Seed may remain viable in the soil for over ten years, making it a difficult plant to control.

Current Management Approaches: Mechanical, biological and chemical methods are some of the effective methods available for control of musk thistle. Hand pulling is most effective on small populations and can be done throughout the year, but is most effective prior to the development of seeds.

Flowers and seedheads should be bagged and disposed of in a landfill to prevent or minimize seed dispersal. Minimizing disturbance to the soil during removal activities will help reduce the chance of germination of seeds stored in the soil.

Two weevils have been introduced from Europe and released in the United States as a biological control for musk thistle, the thistlehead-feeding weevil (Rhinocyllus conicus) and the rosette weevil (Trichosirocalus horridus). These weevils have been released in a number of western states with some notable successes achieved. However, recent observations of unintentional and unanticipated impacts of the thistlehead-feeding weevil to native thistles, including some rare species, has raised a red flag about its continued use, at least in the western U.S. Foliar spraying is effective on established populations of musk thistle.

Apply a 2% solution of glyphosate (e.g., Roundup) or triclopyr (e.g., Garlon) and water plus a 0.5% non-ionic surfactant wetting all leaves and stems. Chlorpyralid (e.g. Transline) is effective at a concentration of 0.5% and is selective to Aster, Buckwheat, and Pea families. A low pressure and coarse spray pattern will limit drift and damage to non-target species. Treatments should be applied during the rosette stage or prior to flowering. Glyphosate is a non-selective systemic (i.e., moves through the plant) herbicide that can kill non-target plants that are only partially contacted by spray. Triclopyr is selective to broadleaf species and is a better choice if native grasses are present.

Suggested Alternative Plants: Although not a popular ornamental in the U.S., suitable native alternatives for musk thistle in the eastern U.S. might include butterfly weed (Asclepias tuberosa), Joe-pye weed (Eupatorium dubium), black-eyed Susan (Rudbeckia fulgida), ironweed (Vernonia noveboracensis), wild blue phlox (Phlox divaricata) and many others. Many plants native to the West are also available. Check with the native plant society in your state for more suggestions.

ASIATIC SAND SEDGE CAREX KOBOMUGI OHWI

Native Range: Coastal areas of northeastern Asia

Description: Asiatic sand sedge is a perennial member of the sedge family (Cyperaceae) adapted to coastal beaches and dunes and possibly the only Carex species found in upper beach habitat along the U.S. Atlantic Coast. The mature sedge is a coarse and stout member of the genus that forms extensive colonies through cord-like rhizomes that extend many feet under the sand and produce new shoots. Flowering and fruiting occurs April through June and individual plants have either male or female flowers. As with many other members of the genus Carex, the flowers are numerous, subtended by scales, and arranged in spikes at the end of a flowering stalk that is triangular in cross section. A papery sac or perigynium encloses the female flowers, each of which develops into a single-seeded fruit, called an achene.

Because flowering culms are evident for a relatively brief period during the spring, and some colonies and new infestations may spread extensively without flowering, it is useful to learn to recognize the plant in its sterile form. Asiatic sand sedge may be confused with at least two colonial, rhizomatous native grass species-American beach grass (Ammophila breviligulata) and beach panic grass (Panicum amarum). Leaves of Asiatic sand sedge are longer tapering than those of the above grasses, have a yellow-green rather than bluish-green cast, and small teeth along the margin that are easily felt or seen with the help of magnification. These differences become more obvious when observed in the field. Several species of another sedge genus, Cyperus, sometimes grow on dunes and on wash flats and strongly resemble Carex when not in flower. However, these

Cyperus species flower from late summer to fall, have leaves without serrated margins and, unlike Asiatic sand sedge, are weakly to nonrhizomatous.

Ecological Threat: Asiatic sand sedge invades wash flat habitat occupied by the federally listed plant, seabeach amaranth (Amaranthus pumilus), which is a poor competitor against it. On established, vegetated sand dunes, Asiatic sand sedge can out-compete native dunebinding grasses, like American beach grass and sea oats (Uniola paniculata). Dunes dominated by Asiatic sand sedge are also more vulnerable to wind blowouts and storm erosion. There is evidence to suggest that fewer native plant species, and fewer individuals, occur on dunes dominated by Asiatic sand sedge than on comparable dunes dominated by the native American beach grass.

Distribution in the United States: Asiatic sand sedge occurs in maritime areas from Massachusetts to North Carolina [DE, MA, MD, NC, NJ, NY, RI, VA].

Habitat in the United States: Asiatic sand sedge grows on primary dunes and on upper parts of ocean beach wash flats that have recently been disturbed by ocean storms. Like American beach grass, it appears to create more habitat for itself by trapping wind-blown sand to form dunes. Sand burial appears to stimulate the growth of rhizomes.

Background: Asiatic sand sedge was first observed in the United States, at Island Beach, New Jersey in 1929. Specimens were collected on the Virginia part of the Delmarva (Delaware-Maryland-Virginia) Peninsula as early as the 1940s. Although the circumstances of its introduction are unclear, sand sedge was apparently introduced intentionally for use as a sand binder in erosion-prone areas and may have spread accidentally as a result of its use as a packing material in ship cargo.

Biology and Spread: Once established, Asiatic sand sedge spreads primarily by vegetative means, through production of rhizomes. Sexual reproduction, which requires both male and female plants to be present, is not necessary for a colony to expand locally. Expansion of a colony was observed at Island Beach, New Jersey, despite the absence of any seedlings. Longdistance dispersal of Asiatic sand sedge is uncertain but it is likely that its seeds are tolerant of salt water immersion and carried by ocean currents and storm surges. Plant fragments may be dispersed by ocean currents, and may remain viable after extended salt-water immersion, but this has not been confirmed. Some observation suggests that inundation by storm surges can kill growing plants. In newly forming colonies, sexual reproduction may be somewhat limited, since plants of the opposite sex may not occur nearby. Pollen may be carried long distances by the wind. Much research is needed to gain a better understanding of modes of dispersal and establishment of Asiatic sand sedge.

Current Management Approaches: Various mechanical and chemical methods have been used successfully in managing Asiatic sand sedge. Regardless of method, it is important to avoid breaking underground parts and leaving them untreated, and to conduct follow-up monitoring and treatment if needed. Mapping infestations with a Global Positioning System (GPS) prior to treatment is very helpful for relocating sites, especially in sandy natural areas like beaches with few permanent landmarks. Cooperation and coordination among coastal area land managers should lead to more effective control.

Because Asiatic sand sedge is capable of forming extensive colonies, early detection and treatment of infestations is critical for effective management. The potential for considerable longdistance dispersal of seeds necessitates routine monitoring and possible follow up treatments, even after it is believed to be eradicated. Because of the likelihood of leaving viable below-ground parts after an excavation, it is important to revisit the site in subsequent years to ensure that an infestation has been eradicated. Manual. Excavation of individual plants by digging and hand-pulling is feasible and has been successful when used to control small infestations (e.g., fewer than 200 shoots). This method may not be economically or logistically feasible on larger control projects. Excavation generally involves digging with a shovel under and around each individual plant shoot to expose and loosen the roots. Individual shoots are often connected to other shoots by cordlike rhizomes that are about ¼ inch thick and often of considerable length.

Once shoot and roots are loose, all rhizomes need to be gently excavated by hand, following them through the sand to minimize breaking. Rhizome parts left buried are likely to grow into new plants. Because the tips of new tillers (shoots) can be sharp enough to puncture skin, it is important to wear thick gloves when handling below-ground parts. Plants should be removed from the beach and disposed of in habitat unsuitable for the sedge (e.g., lawns), spread out to dry, or composted in black plastic until dead. Chemical. Larger colonies of Asiatic sand sedge that have formed considerable dunes are probably most effectively controlled using chemical herbicides. A 2% glyphosate (e.g. Roundup, Rodeo, etc.) and water solution applied to the leaves during the growing season has provided effective control. One or two treatments in the same season followed by spot treatments are usually needed. Mid-summer (June through July in Maryland and New Jersey) treatments are just as effective as fall (October in Maryland) applications and allow for same season monitoring and re-treatment. Because rhizomes can be extensive, follow-up monitoring and treatment are necessary for several seasons to ensure long-term control.

Good coverage of herbicide is needed but can be difficult because of the plant's narrow leaves. To help track application and to minimize misapplication and waste, a colorant can be added to the spray solution. Herbicide applications should be made when the chance of rain is low for at least six hours after application and when winds are minimal (e.g., 0-7 mph), to minimize drift of herbicide to non-target areas. Herbicide users should read and follow all label instructions and, when possible, mix chemicals where a spill containment and/or clean-up facility is available instead of on site. Transport of herbicide is likely to be more rapid through sand than in other soils, and microbial activity that can break down herbicides is likely to be low in beach sand. When it is necessary to mix herbicide on the beach or dunes, it is recommended to mix over a waterproof basin set on top of a waterproof tarp.

Suggested Alternative Plants: Asiatic sand sedge was originally introduced as a dune stabilizer, although it is apparently less effective in this role than native species, such as American beach grass (Ammophila breviligulata), which occurs throughout the North American range of Asiatic sand sedge. In the southernmost part of this range, sea oats (Uniola paniculata) is the dominant native dune binding grass.

Dune Restoration and Planting. Once successful control of Asiatic sand sedge has been achieved, establishing native vegetation is an integral part of dune restoration. Native species such as American beachgrass (Ammophila breviligulata) and sea oats (Uniola paniculata) should be planted to protect vulnerable dunes from storm damage and blowouts and to prevent recolonization by Asiatic sand sedge. American beachgrass establishes itself well on primary foredunes were shifting sands are common and should be planted during late winter to early spring. In primary backdune areas and places where sands are usually more stable, consideration should be given to planting species such as seaside goldenrod (Solidago sempervirens), beach panic grass (Panicum amarum), dune panic grass (Panicum amarulum), and sea-rocket (Cakile edentula), in combination with American beachgrass and sea oats.

SPOTTED KNAPWEED CENTAUREA BIEBERSTEINII DC.

Native Range: Central Europe, east to central Russia, Caucasia, and western Siberia.

Description: Spotted knapweed is a biennial or shortlived perennial member of the Sunflower (Asteraceae) family. Its name is derived from the spots formed by black margins on the flower bract tips. Spotted knapweed typically forms a basal rosette of leaves in its first year and flowers in subsequent years. Rosette leaves are approximately 8 inches long by 2 inches wide, borne on short stalks, and deeply lobed once or twice on both sides of the center vein, with lobes oblong and wider toward the tip. The taproot is stout and deep. Flowering stems are erect, 8 to 50 inches tall, branched above the middle, and sparsely to densely hairy. Stem leaves alternate along the stem, are unstalked, and may be slightly lobed, or linear and unlobed. Leaf size decreases towards the tip of the stem.

Flowers are purple to pink, rarely white, with 25 to 35 flowers per head. Plants bloom from June to October, and flower heads usually remain on the plant. Flower heads are oblong or oval shaped, ¼ inch wide and ½ inch across, and are single or borne in clusters of two or three at the branch ends. Leaf-like bracts surrounding the base of the flower head are oval and yellow green, becoming brown near the base. The margins of these bracts have a soft spine-like fringe, with the center spine being shorter than the lateral spines. The brown, oval seeds are 1/16 to 1/8 inch long, with pale longitudinal lines and a short fringe on one end.

Ecological Threat: Spotted knapweed infests a variety of natural and semi-natural habitats including barrens, fields, forests, prairies, meadows, pastures, and rangelands. It outcompetes native plant species, reduces native plant and animal biodiversity, and decreases forage production for livestock and wildlife. Spotted knapweed may degrade soil and water resources by increasing erosion, surface runoff, and stream sedimentation. It has increased at an estimated rate of 27% per year since 1920 and has the potential to invade about half of all the rangeland (35 million acres) in Montana alone.

Distribution in the United States: Spotted knapweed is a widely distributed species reported to occur throughout Canada and in every state in the U.S. except Alaska, Georgia, Mississippi, Oklahoma and Texas (see map).. It has been designated as a noxious weed in Arizona, California, Colorado, Idaho, Minnesota, Montana, Nebraska, Nevada, New Mexico, North Dakota, Oregon, South Dakota, Utah, Washington, and Wyoming. It has been identified as invasive in natural areas by eighteen organizations in twenty-six states (Arizona, California, Colorado, Connecticut, Delaware, Kentucky, Idaho, Illinois, Massachusetts, Maryland, Michigan, Minnesota, Montana, North Carolina, New Jersey, New York, Oregon, Pennsylvania, South Dakota, Tennessee, Utah, Virginia, Washington, Wisconsin, West Virginia, and Wyoming). Fifteen national parks also identify spotted knapweed as an invasive plant and a threat to natural habitats.

Habitat in the United States: Spotted knapweed is found at elevations up to and over 10,000 feet and in precipitation zones receiving 8 to 80 inches of rain annually. Spotted knapweed prefers well-drained, light-textured soils that receive summer rainfall, including open forests dominated by ponderosa pine and Douglas fir, and prairie habitats dominated by Idaho fescue, bluebunch wheatgrass, and needle-and-thread grass. Disturbance allows for rapid establishment and spread; however, spotted knapweed is capable of invading well managed rangelands. Spotted knapweed does not compete well with vigorously growing grass in moist areas. In seasonally dry areas, spotted knapweed's taproot allows it to access water from deep in the soil, beyond the reach of more shallowly rooted species.

Background: Spotted knapweed was introduced to North America from Eurasia as a contaminant in alfalfa and possibly clover seed, and through discarded soil used as ship ballast. It was first recorded in Victoria, British Columbia in 1883 and spread further in domestic alfalfa seeds and hay before it was recognized as a serious problem.

Biology and Spread: Spotted knapweed plants in North America generally live 3 to 7 years but can live up to nine years or longer. Plants re-grow from buds on the root crown. Reproduction is by seed, and plants are capable of producing 500-4,000 seeds per square foot per year. About 90% of the seeds are viable at the time of dispersal, and they can remain viable in the soil for 5-8 years. Most seeds are dispersed near the parent plant but can be transported by people, wildlife, livestock, vehicles, and in soil, crop seed, and contaminated hay. Gravel pits, soil stockpiles, powerlines, grain elevators, railroad and equipment yards are important seed distribution points.

Current Management Approaches: The most cost effective management strategy for spotted knapweed is to prevent its spread to non-infested areas. Spread by seed can be minimized by avoiding travel through infested areas; by cleaning footwear, clothing, backpacks, and other items after hiking through infested areas; by not grazing livestock when ripe seeds are present in the flower heads; and by using weed-free hay. Manual and Mechanical. Small infestations of spotted knapweed can be controlled by persistent hand-pulling done prior to seed set. Gloves should be worn because of the possibility of skin irritation. Because spotted knapweed can re-grow from the base, care must be taken to remove the entire crown and taproot.

Biological Control: A variety of natural enemies are used as biological control agents for large infestations of spotted knapweed. Most biocontrol techniques use insect larvae to damage the root, stem, leaf, or flower. Two species of seed head flies, Urophora affinis and U. quadrifasciata, are well-established on spotted knapweed. The larvae of these species reduce seed production by as much as 50% by feeding on spotted knapweed seed heads and causing the plant to form galls. Three moth species (Agapeta zoegana, Pelochrista medullana, and Pterolonche inspersa) and a weevil (Cyphocleonus achates) that feed on spotted knapweed roots have also been released.

The collective stress on the plant caused by these insects reduces seed production and may lead to reduced competitiveness. Biological control agents may be more effective when combined with other control methods such as herbicides, grazing, and revegetation with desirable, competitive plants.

Chemical: Control of spotted knapweed infestations using three chemical herbicides (2,4-D, clopyralid, and picloram) has been reported but is problematic. Existing plants can be killed with 2,4-D but it needs to be reapplied yearly to control new plants germinating from seed stored in the soil. Picloram is a more persistent herbicide and has controlled knapweed for three to five years when applied at 0.25 lb/acre at any stage of plant growth; or with clopyralid (0.24 lb/acre) or clopyralid (0.2 lb/acre) plus 2,4-D (1 lb./acre) applied during bolt or bud growth stage. In the absence of desirable native grasses, longevity of control may be increased by revegetating with competitive grasses and forbs. Picloram may pose a risk of groundwater contamination where soils are permeable, particularly where the water table is shallow. Other methods. Long-term grazing by sheep and goats has been found to control spotted knapweed. Burning, cultivation, and fertilization typically are not effective on spotted knapweed unless combined with other methods of control.

CANADA THISTLE CIRSIUM ARVENSE (L.) SCOP.

Native Range: Temperate regions of Eurasia

Description: Canada thistle is an herbaceous perennial in the aster family with erect stems 1½-4 feet tall, prickly leaves and an extensive creeping rootstock. Stems are branched, often slightly hairy, and ridged. Leaves are lance-shaped, irregularly lobed with spiny, toothed margins and are borne singly and alternately along the stem. Rose-purple, lavender, or sometimes white flower heads appear from June through October, generally, and occur in rounded, umbrellashaped clusters.

The small, dry, single-seeded fruits of Canada thistle, called achenes, are 1-1½ inches long and have a feathery structure attached to the seed base. Many native species of thistle occur in the U.S., some of which are rare. Because of the possibility of confusion with native species, Canada thistle should be accurately identified before any control is attempted.

Ecological Threat: Natural communities that are threatened by Canada thistle include nonforested plant communities such as prairies, barrens, savannas, glades, sand dunes, fields and meadows that have been impacted by disturbance. As it establishes itself in an area, Canada thistle crowds out and replaces native plants, changes the structure and species composition of natural plant communities and reduces plant and animal diversity.

This highly invasive thistle prevents the coexistence of other plant species through shading, competition for soil resources and possibly through the release of chemical toxins poisonous to other plants. Canada thistle is declared a "noxious weed" throughout the U.S. and has long been recognized as a major agricultural pest, costing tens of millions of dollars in direct crop losses annually and additional millions in costs for control. Only recently have the harmful impacts of Canada thistle to native species and natural ecosystems received notable attention.

Distribution in the United States: Canada thistle is distributed throughout the northern U.S., from northern California to Maine and southward to Virginia. It is also found in Canada, for which it was named. Canada thistle has been identified as a management problem in many national parks and in preserves of The Nature Conservancy in the upper Midwest, Plains states, and the Pacific northwest.

Habitat in the United States: Canada thistle grows in barrens, glades, meadows, prairies, fields, pastures, and waste places. It does best in disturbed upland areas but also invades wet areas with fluctuating water levels such as streambank sedge meadows and wet prairies.

Background: Canada thistle was introduced to the United States, probably by accident, in the early 1600s and, by 1954, had been declared a noxious weed in forty three states. In Canada and the U.S., it is considered one of the most tenacious and economically important agricultural weeds, but only in recent years has it been recognized as a problem in natural areas.

Biology and Spread: Canada thistle produces an abundance of bristly-plumed seeds which are easily dispersed by the wind. Most of the seeds germinate within a year, but some may remain viable in the soil for up to twenty years or more.

Vegetative reproduction in Canada thistle is aided by a fibrous taproot capable of sending out lateral roots as deep as 3 feet below ground, and from which shoots sprout up at frequent intervals. It also readily regenerates from root fragments less than an inch in length.

Current Management Approaches: Management of Canada thistle can be achieved through hand-cutting, mowing, controlled burning, and chemical means, depending on the level of infestation and the type of area being managed.

Due to its perennial nature, entire plants must be killed in order to prevent regrowth from rootstock. Hand-cutting of individual plants or mowing of larger infestations should be conducted prior to seed set and must be repeated until the starch reserves in the roots are exhausted. Because early season burning of Canada thistle can stimulate its growth and flowering, controlled burns should be carried out late in the growing season for best effects.

In natural areas where Canada thistle is interspersed with desirable native plants, targeted application of a systemic herbicide such as glyphosate (e.g., Roundup or Rodeo), which carries plant toxins to the roots, may be effective.

For extensive infestations in disturbed areas with little desirable vegetation, broad application of this type herbicide may be the most effective method. Repeated applications are usually necessary due to the long life of seeds stored in the soil.

LEAFY SPURGE EUPHORBIA ESULA L.

Native Range: Europe and Asia

Description: Leafy spurge is a member of the spurge family, or Euphorbiaceae, characterized by plants containing a white milky sap and flower parts in three's. Leafy spurge is an erect, branching, perennial herb 2 to 3½ feet tall, with smooth stems and showy yellow flower bracts. Stems frequently occur in clusters from a vertical root that can extend many feet underground. The leaves are small, oval to lance-shaped, somewhat frosted and slightly wavy along the margin. The flowers of leafy spurge are very small and are borne in greenish-yellow structures surrounded by yellow bracts. Clusters of these showy, yellow bracts open in late May or early June, while the actual flowers do not develop until mid-June.

Ecological Threat: Leafy spurge displaces native vegetation in prairie habitats and fields through shading and by usurping available water and nutrients and through plant toxins that prevent the growth of other plants underneath it. Leafy spurge is an aggressive invader and, once present, can completely overtake large areas of open land.

Distribution in the United States: Leafy spurge occurs across much of the northern U.S., with the most extensive infestations reported for Montana, North Dakota, Nebraska, South Dakota, and Wyoming. It has been identified as a serious pest in a number of national parks and in preserves of The Nature Conservancy in eleven northern states.

Habitat in the United States: Leafy spurge tolerates moist to dry soil conditions but is most aggressive under dry conditions where competition from native plants is reduced. It is capable of invading disturbed sites, including prairies, savannas, pastures, abandoned fields and roadside areas.

Background: Leafy spurge was transported to the U.S. possibly as a seed impurity in the early 1800s. First recorded in Massachusetts in 1827, leafy spurge spread quickly and reached North Dakota within about 80 years.

Biology and Spread: Leafy spurge reproduces readily by seeds that have a high germination rate and may remain viable in the soil for at least seven years, enhancing its chances of recovery over time. Its seed capsules open explosively, dispersing seed up to 15 feet from the parent plant and may be carried further by water and wildlife. Leafy spurge also spreads vegetatively at a rate of several feet per year. The root system is complex, can reach 15 or more feet into the ground, and may have numerous buds.

Current Management Approaches: Because of its persistent nature and ability to regenerate from small pieces of root, leafy spurge is extremely difficult to eradicate. Biological control offers a highly promising management tactic for leafy spurge. The U.S. Department of Agriculture has shown success using six natural enemies of leafy spurge imported from Europe. These include a stem and root-boring beetle (Oberea erythrocephala), four root-mining flea beetles (Aphthona spp.) and a shoot-tip gall midge (Spurgia esulae). Large scale field-rearing and release programs are carried out cooperatively by federal and State officials in many northern states. The results are not as immediate as when herbicides are used but, if pesticide use is kept to a minimum, large numbers of these agents build up within a few years and have shown impressive results. Several systemic herbicides have been found to be effective if applied in June, when the flowers and seeds are developing, or in early to mid-September, when the plants are moving nutrients downward into the roots.

Preliminary research suggests that chemical treatment in the fall followed by a spring burn to reduce seed germination may be an effective strategy for reducing leafy spurge infestations. Multiple treatments are necessary every year for several years, making leafy spurge control an extremely expensive undertaking. If left uncontrolled for a single year, leafy spurge can re-infest rapidly. Prescribed burning, in conjunction with herbicides, may also be effective.

COGON GRASS IMPERATA CYLINDRICA (L.) PALISOT

Native Range: Southeast Asia, Philippines, China, and Japan

Description: Cogon grass is a perennial, rhizomatous grass that grows from 2 to over 4 feet in height. The leaves are about an inch wide, have a prominent white midrib, and end in a sharp point. Leaf margins are finely toothed and are embedded with silica crystals. The upper surface of the leaf blade is hairy near the base; the undersurface is usually hairless. The flowers are arranged in a silvery, cylindrical, branching structure, or panicle, about 3-11 inches long and 1½ inches wide.

Ecological Threat: Cogon grass can invade and overtake disturbed ecosystems, forming a dense mat of thatch and leaves that makes it nearly impossible for other plants to coexist. Large infestations of cogon grass can alter the normal fire regime of a fire-driven ecosystem by causing more frequent and intense fires that injure or destroy native plants. Cogon grass displaces a large variety of native plant species used by native animals (e.g., insects, mammals, and birds) as forage, host plants and shelter. Some ground-nesting species have also been known to be displaced due to the dense cover that cogon grass creates.

Distribution in the United States: Cogon grass is distributed throughout the south and southeastern United States as far west as eastern Texas. There have been reports of cogon grass surviving as far north as Virginia, West Virginia and Maryland.

Habitat in the United States: Cogon grass is a hardy species, tolerant of shade, high salinity and drought. It can be found in virtually any ecosystem, especially those experiencing disturbance. Cogon grass has been found growing on sand dunes in the southeast, along roadsides, forests, open fields, and up to the edge of standing water.

Background: Cogon grass was introduced to the United States both accidentally and intentionally. Cogon grass was first introduced to the U.S. in Mobile, Alabama, via shipping crates that contained cogon grass as a packing material. It was also brought in and distributed by the U.S.D.A. for use as a forage grass and for soil erosion control. Cogon grass is also sold by the nursery trade as an ornamental grass, valued for its attractive foliage and hardiness.

Biology and Spread: Cogon grass reproduces both vegetatively and from seed. A single plant can produce several thousand very small seeds that may be carried great distances by the wind. Vegetative spread of cogon grass is aided by its tough and massive rhizomes that may remain dormant for extended periods of time before sprouting. Rhizomes of cogon grass may be transported to new sites in contaminated fill dirt or by equipment used in infested areas.

Current Management Approaches: An integrated approach to cogon grass management, including chemical, mechanical and cultural methods, is necessary to achieve successful, long lasting control. Effective management of cogon grass has been achieved by the following combined mechanical-chemical protocol. First, the infested area is mowed in late spring to remove last year's growth and the accumulated thatch layer. About six to eight weeks later, when about eighty percent of the cogon grass has re-sprouted to a height of 6-12 inches, the site needs to be disced as deeply as possible. (Discing may not be possible in all areas, due to the sensitive nature of some ecosystems). When adequate regrowth of the cogon grass has occurred, systemic herbicides (chemicals carried through the plant tissues to the roots) are applied.

The best time to apply herbicides is in the early fall before first frost. A 2% solution of glyphosate (e.g., Roundup) is recommended in areas that will be immediately revegetated, because glyphosate has no residual soil activity. In areas where immediate revegetation is not planned, and non-target plant damage is not a concern, application of a 1-1.5% solution of imazapyr (e.g., Arsenal) may be considered.

Important: Because imazapyr is highly active in soil, it has a high potential for leaching into groundwater, so should only be used when groundwater impacts can be eliminated or minimized. Additionally, nearby trees or other plants may be damaged by improper application of this herbicide.

Revegetation may be necessary following herbicide treatment, to prevent soil erosion and to help reduce re-infestation by cogon grass. For roadside areas, revegetation with bahia grass and bermuda grass have been used successfully for these purposes. In natural areas, the choice of which species to use for revegetation becomes more difficult. For some areas, assisting the process of natural revegetation succession may be the best choice. Once decisions are made regarding the ultimate goal of the restcration project, revegetation plans should be made accordingly. Regardless of the goal, the area should be revegetated quickly to allow a stable plant community to be established. Lastly, it is important to exercise diligence, as it will be necessary to scout areas that have been treated for cogon grass and spot treat new plants with herbicides.

It may not be possible to use all of the methods prescribed above in every situation, but most effective control will be gained by using as many of the steps outlined above as possible. In areas where burning, mowing, or discing, are not possible, spot treatment with herbicides will help to control cogon grass. Revisit treated areas frequently and retreat with herbicides as necessary. The best time to begin a control program is late spring to mid-summer when cogon grass is experiencing peak growth. Some control measures, such as mowing and spot spraying with herbicides, can be implemented year round. Burning has also been used successfully in controlling cogon grass. As with mowing, burning stimulates the growth and spread of cogon grass, making follow-up control a necessity. If you are interested in attempting this technique, contact management specialists listed below for more specific information. Also, be sure to obtain all required permits before attempting a burn, even on small infestations.

CHINESE LESPEDEZA LESPEDEZA CUNEATA (DUMONT) G. DON

Native Range: Eastern Asia

Description: Chinese lespedeza is a warm season, perennial herb in the pea family, or Fabaceae. It has an erect growth form, ranging from about 3 to 5½ feet in height, and leaves that alternate along the stem. Each leaf is divided into three smaller leaflets, about ½ to 1 inch long, which are narrowly oblong and pointed, with awlshaped spines. Leaflets are covered with densely flattened hairs, giving a grayishgreen or silvery appearance. Mature stems are somewhat woody and fibrous with sharp, stiff, flattened bristles. Violet to purple flowers emerge either singly or in clusters of 2-4, from the axils of the upper and median leaves.

Ecological Threat: Chinese lespedeza, sometimes called sericea lespedeza, is primarily a threat to open areas such as meadows, prairies, open woodlands, wetland borders and fields.

Once it gains a foothold, it can crowd out native plants and develop an extensive seed bank in the soil, ensuring its long residence at a site. Established dense stands of lespedeza suppress native flora and its high tannin content makes it unpalatable to native wildlife as well as livestock.

Distribution in the United States: Chinese lespedeza is now found throughout the U.S.

Habitat in the United States: Chinese lespedeza can grow in a variety of habitats including severely eroded sterile soils. It will invade open woodlands, fields, prairies, borders of ponds and swamps, meadows, and open disturbed ground, but is intolerant of shade.

Background: Chinese lespedeza is native to eastern Asia and was first introduced in the southern United States. Widespread use of lespedeza by federal and state agencies for bank stabilization, and soil improvement; wildlife, forage and cover; and hay facilitated its spread throughout the eastern United States.

Biology and Spread: Chinese lespedeza begins growth from root crown buds at the base of last year's stem. The flowers begin to develop in late July and continue through October. Within the Lespedeza genus there are no specialized structures for seed dispersal. Dispersal is aided by animals consuming the fruits and passing the seeds. A study on natural populations found that several species of Lespedeza comprise 1.5% to 86.8% of the annual diet of bobwhite quail in the southeastern U.S. Autumn dispersal is aided by the haying of infested fields. Scarification is necessary for the germination of lespedeza seeds. Mature seeds of this genus remain viable for up to twenty years; one study found a germination rate of 60% after cold storage for 55 years. Seedlings may represent only 1% of the seeds actually available in the soil.

Current Management Approaches: Mechanical and chemical methods are the most effective options currently available for Chinese lespedeza. Hand pulling is impractical due to lespedza's extensive perennial root system. Mowing plants in the flower bud stage for two or three consecutive years may reduce the vigor of lespedeza stands and control further spread. Plants should be cut as low to the ground as possible and impact to adjacent native plants should be minimized as much as possible. Since root reserves increase up to the flower bud stage, all herbicide treatments should be completed in early to mid summer.

The addition of a non-ionic surfactant at a concentration of 0.5% improves the effectiveness of foliar treatments. Triclopyr and clopyralid have been shown to be effective in controlling Chinese lespedeza. A 2% solution Triclopyr or 0.5% solution of clopyralid thoroughly mixed with water is effective during the vegetative stage prior to branching or during flowering. Treatments should cover the leaves and stems of plants to the point of runoff. These herbicides are not labeled for use in wet areas or adjacent to streams. On wet sites a 2% solution of glyphosate is effective from last June until seed set.

PURPLE LOOSESTRIFE LYTHRUM SALICARIA L.

Native Range: Eurasia; throughout Great Britain, and across central and southern Europe to central Russia, Japan, Manchuria China, southeast Asia and northern India

Description: Purple loosestrife is an erect perennial herb in the loosestrife family, with a square, woody stem and opposite or whorled leaves. Leaves are lance-shaped, stalkless, and heart-shaped or rounded at the base. Plants are usually covered by a downy pubescence. Loosestrife plants grow from four to ten feet high, depending upon conditions, and produce a showy display of magenta-colored flower spikes throughout much of the summer. Flowers have five to seven petals. Mature plants can have from 30 to 50 stems arising from a single rootstock.

Ecological Threat: Purple loosestrife adapts readily to natural and disturbed wetlands. As it establishes and expands, it outcompetes and replaces native grasses, sedges, and other flowering plants that provide a higher quality source of nutrition for wildlife. The highly invasive nature of purple loosestrife allows it to form dense, homogeneous stands that restrict native wetland plant species, including some federally endangered orchids, and reduce habitat for waterfowl.

Distribution in the United States: According to the U.S. Fish and Wildlife Service, purple loosestrife now occurs in every state except Florida.

Habitat in the United States: Purple loosestrife is capable of invading many wetland types, including freshwater wet meadows, tidal and non-tidal marshes, river and stream banks, pond edges, reservoirs, and ditches.

Background: Purple loosestrife was introduced to the northeastern U.S. and Canada in the 1800s, for ornamental and medicinal uses. It is still widely sold as an ornamental, except in states such as Minnesota, Wisconsin, and Illinois where regulations now prohibit its sale, purchase and distribution.

Biology and Spread: Purple loosestrife enjoys an extended flowering season, generally from June to September, which allows it to produce vast quantities of seed. The flowers require pollination by insects, for which it supplies an abundant source of nectar. A mature plant may have as many as thirty flowering stems capable of producing an estimated two to three million minute seeds per year.

Purple loosestrife also readily reproduces vegetatively through underground stems at a rate of about one foot per year. Many new stems may emerge vegetatively from a single rootstock of the previous year. "Guaranteed sterile" cultivars of purple loosestrife are actually highly fertile and able to cross freely with purple loosestrife and with other native Lythrum species. Therefore, outside of its native range, purple loosestrife of any form should be avoided.

Current Management Approaches: Small infestations of young purple loosestrife plants may be pulled by hand, preferably before seed set. For older plants, spot treating with a glyphosate type herbicide (e.g., Rodeo for wetlands, Roundup for uplands) is recommended. These herbicides may be most effective when applied late in the season when plant are preparing for dormancy. However, it may be best to do a mid-summer and a late season treatment, to reduce the amount of seed produced.

While herbicides and hand removal may be useful for controlling individual plants or small populations, biological control is seen as the most likely candidate for effective long term control of large infestations of purple loosestrife. As of 1997, three insect species from Europe have been approved by the U.S. Department of Agriculture for use as biological control agents. These plant-eating insects include a root-mining weevil (Hylobius transversovittatus), and two leaf-feeding beetles (Galerucella calmariensis and Galerucella pusilla). Two flower-feeding beetles (Nanophyes) that feed on various parts of purple loosestrife plants are still under investigation. Galerucella and Hylobius have been released experimentally in natural areas in 16 northern states, from Oregon to New York. Although these beetles have been observed occasionally feeding on native plant species, their potential impact to non-target species is considered to be low.

JAPANESE STILT GRASS MICROSTEGIUM VIMINEUM (TRIN.) CAMUS

Native Range: Japan, Korea, China, Malaysia and India

Description: Japanese stilt grass, also known as Asian stilt grass, Vietnamese stilt grass, Nepal microstegium, and Chinese packing grass, is an annual grass (family Poaceae) with a sprawling habit that may grow to 3 feet in height. Its thin, pale green, lance-shaped leaves, about 3 inches in length, alternate along a branched stalk and have a silvery stripe of reflective hairs down the middle of the upper leaf surface. Delicate spikes of flowers emerge from slender tips beginning in late summer and continuing into the fall. Seeds may persist through the fall.

Ecological Threat: Japanese stilt grass is adapted to low light conditions and threatens native understory vegetation in open to shady locations. It spreads opportunistically following soil disturbance to form dense patches, displacing native wetland and forest vegetation as the patch expands.

Distribution in the United States: Japanese stilt grass is currently established in sixteen eastern states, from New York to Florida.

Habitat in the United States: Stilt grass occurs on stream banks, river bluffs, floodplains, emergent and forested wetlands, moist woodlands, early successional fields, uplands, thickets, roadside ditches, gas and power line corridors and home lawns and gardens. It readily invades and is most common in disturbed shaded areas like floodplains that are prone to natural scouring, and areas subject to mowing, tilling and other soil disturbing activities. Japanese stilt grass appears to be associated primarily with moist, acidic to neutral soils that are high in nitrogen. It occurs opportunistically in areas of open soil that are generally not already occupied by other species.

Background: Introduced into Tennessee around 1919, Japanese stilt grass may have accidentally escaped as a result of its use as a packing material for porcelain.

Biology and Spread: Japanese stilt grass is a colonial species that spreads by rooting at nodes along the stem. A new plant emerges from each node. It also spreads by seed and each plant can produce an estimated 100-1,000 seeds. Once established at a site, seed stored in the soil will ensure regrowth for several to many years. Studies have shown that stilt grass seed remains viable in the soil for at least three years and germinates readily following soil disturbance. Although seed dispersal of stilt grass is not fully understood, seeds are probably carried by water currents in streamside habitats and floods, and transported widely in hay and soil.

Current Management Approaches: Whenever possible, prevent the introduction of Japanese stilt grass from invaded sites into adjacent natural plant communities by avoiding disturbance to vegetation and soil in these areas. Early control of new infestations will also greatly reduce the likelihood of its establishment.

Because it is shallow-rooted, stilt grass may be pulled by hand at any time, but if done early in the summer, disturbance to the soil may allow for germination of stored stilt grass seed. Hand pulling of plants will need to be repeated and continued for many seasons. A more effective mechanical method might be to wait until late summer (September) when the plants are in peak bloom but before seed is produced, and simply cut them back using a mower or "weed whacker". Being an annual plant, Japanese stilt grass cut late in the season will die back for the winter and not produce additional vegetative shoots.

For extensive infestations, where mechanical methods are not feasible, a systemic herbicide like glyphosate (e.g., Roundup), an herbicidal soap that kills the plants back (e.g., Scythe) and herbicides specific to annual grasses may be a more effective choice. If applying glyphosate to stilt grass in wetland sites, use the formulation labeled for wetland areas (e.g., Rodeo). No biological controls are currently available for this plant.

EURASIAN WATERMILFOIL MYRIOPHYLLUM SPICATUM L.

Native Range: Eurasia and Africa

Description: Eurasian watermilfoil, also called spike watermilfoil, is an emergent, herbaceous aquatic plant in the Water-milfoil family, or Haloragaceae. Stems grow to the water surface, usually extending 3 to 10, but as much as 33, feet in length and frequently forming dense mats. Stems of Eurasian milfoil are long, slender, branching, hairless, and become leafless toward the base. New plants may emerge from each node (joint) on a stem, and root upon contact with mud. The grayish-green leaves of Eurasian watermilfoil are finely divided and occur in whorls of three or four along the stem, with 12-16 pairs of fine, thin leaflets about 12 inches long.

These leaflets give milfoil a feathery appearance that is a distinguishing feature of the plant. Eurasian watermilfoil produces small yellow, 4-parted flowers on a spike that projects 2-4 inches above the water surface. The fruit is a hard, segmented capsule containing four seeds.

Ecological Threat: Eurasian milfoil can form large, floating mats of vegetation on the surface of lakes, rivers, and other water bodies, preventing light penetration for native aquatic plants and impeding water traffic. The plant thrives in areas that have been subjected to various kinds of natural and manmade disturbance.

Distribution in the United States: Watermilfoil occurs in thirty-three states east of the Mississippi River and has recently been found in Colorado. It is abundant in the Chesapeake Bay, the tidal Potomac River, and several Tennessee Valley reservoirs.

Habitat in the United States: Typical habitat for Eurasian watermilfoil includes fresh to brackish water of fish ponds, lakes, slow-moving streams, reservoirs, estuaries, and canals.. It is tolerant of many water pollutants. Eurasian watermilfoil tends to invade disturbed areas where native plants cannot adapt to the alteration. It does not spread rapidly into undisturbed areas where native plants are well established. By altering waterways, humans have created a new and unnatural niche where milfoil thrives.

Background: Eurasian watermilfoil was accidentally introduced from Eurasia in the 1940s. Two theories exist as to how it entered North America: (1) it escaped from an aquarium, or (2) it was brought in attached to commercial or private boats. A resort owner is thought to have introduced watermilfoil into the Tennessee Valley Authority reservoir system in 1953.

Biology and Spread: Most regeneration of Eurasian watermilfoil is from rhizomes, fragmented stems, and axillary buds that develop throughout the year. Flower spikes often remain above water until pollination is complete, then re-submerge. Although seeds are usually viable, they are not an important means of dispersal.

Current Management Approaches: Large harvesting equipment can be used to mechanically remove milfoil in larger areas; a sturdy hand-rake can be used for smaller areas. Other available options include manipulation of water level, light penetration and chemical control. Potential impacts to existing native aquatic plant species should be evaluated carefully before implementing any of these techniques. For the single harvest, removal should take place just before peak biomass is obtained (early summer). Substantial regrowth may occur if this is done too early. Better results appear with multiple harvests in the same growing season. If multiple harvests are not possible, then sustaining annual harvests is an option. All fragments of milfoil plants must be removed to achieve adequate control.

Where water levels are under manual control, raising or lowering the water level can be an effective way to reduce the growth of milfoil. By raising the water level, plants can be "drowned" by not having access to enough light.

By lowering the water level, plants can be dehydrated and, at the right time of the year, frozen to death. This type of control is best used in conjunction with herbicides and shade barriers. Bankside plantings, floating native plant species, light limiting dyes, or shade barriers are effective ways of reducing the amount of light reaching the plants, and may reduce overall growth rates. Barriers can be used to prevent the movement and spread of aquatic weeds in ponds and lakes. A barrier is usually a suspended blocking screen that hangs vertically from a cable to a depth of about 4 meters; the cable is suspended by drum floats.

Fluridone (the active ingredient in Sonar AS) is a selective herbicide for milfoil and several other exotic aquatic weeds. There are no restrictions on swimming, fishing, or drinking after application, and season-long control can be achieved with one application. Fluridone is available in liquid or granular form, and can be used as a spot treatment or on an entire waterway. For best results, applications should be made before or during the early stages of active growth.

BURMA REED NEYRAUDIA REYNAUDIANA (KUNTH) KENG EX A.S. HITCHC.

Native Range: Southeast Asia and Indomalaya (Japan, southern China, Viet Nam, Laos, Cambodia, Thailand, Malaya, Myanmar (Burma), Bhutan, Nepal, and eastern India)

Description: Burma reed, also known as silk reed, cane grass, and false reed, is a tall, perennial, large-plumed grass (Poaceae family) that grows in clumps in sunny upland areas. Stems, including the flower stalks, are from 3 to 15 feet in height, depending on soil and moisture conditions. The leaves are 8 to 10 inches long and hairless, except for a single line of horizontal hairs at the juncture of the upper and lower portions of the leaf. Stems are approximately ½ inch in width, are round, solid, and have nodes (stem-leaf junctures) every 3 to 5 inches along the stem. The flower plumes, which can be up to 3 feet long, are composed of many hundreds of tiny flowers and have a shimmery, silky appearance. Flowering occurs in April and October, each clump producing an average of forty stalks and twelve to twenty flowering plumes. Burma reed resembles several other tall grasses, including common reed (Phragmites communis), giant reed (Arundo donax), pampas grass (Cortaderia selloana) and sugar cane (Saccharum officinarum).

Ecological Threat: Burma reed damages native ecosystems by crowding and shading out understory plant species and by creating conditions for extremely hot and destructive wildfires. In southern Florida (Miami-Dade County), it is a serious threat to the globally imperiled pine rocklands community whose pine canopy was largely destroyed in 1992 by Hurricane Andrew. Burma reed is a highly combustible fuel source because of its overall plant mass, its large feathery flower plumes, and the dense, hay-like leaf litter it produces.

This hay-like litter enhances the fire's movement along the ground, while the flower plumes carry the flames high into the air. With the aid of winds, these plumes often detach and fly through the air like torches, providing the potential for additional spread. Photographs of its ignition during a wildfire show flames leaping over 30 feet high, threatening nearby tree canopies.

Distribution in the United States: Burma reed is found throughout southern Florida in the counties of Miami-Dade, Broward, Palm Beach, Lee, and Collier, and the Florida Keys.

Habitat in the United States: In its native range, which is characterized by a warm, subtropical climate, Burma reed occurs in bogs, in open savannahs, on upland cliffs, and along forest and road edges, and thrives from sea level to altitudes of 6,500 feet. In the U.S., Burma reed initially colonizes the margins of roadways, fields, and forests, from which it can spread to undisturbed areas. The ability of Burma reed to survive at high altitudes in its native range indicates a tolerance to cold and the potential for it to spread further north in the U.S.

Background: Burma reed was first introduced into the United States in 1916 by the U.S. Department of Agriculture, possibly to investigate its potential as an ornamental plant. It was grown in a test garden in Coconut Grove, Florida, from which it escaped and spread. By 1990, it had become established in the wild as far as 30 miles from Coconut Grove and along disturbed edges throughout Miami-Dade County. Burma reed has no known economic value and, in Bhutan, is reported to be poisonous to buffalo.

Biology and Spread: Burma reed reproduces by seed and through underground stems called rhizomes. Burma reed plants flower twice each year, producing hundreds of thousands of tiny seeds that are dispersed by the wind. New clumps of Burma reed emerge from rhizomes that may be embedded in sand, soil, or rubble. Seeds and rhizomes are also transported inadvertently in limestone rock from infested quarries that is carried by train from Miami-Dade County, Florida to concrete manufacturers throughout the southeastern United States. This unintentional movement of Burma reed material allows it to invade new sites in Florida and adjacent states near limestone distribution centers.

Current Management Approaches: Restoration of sites infested with Burma reed requires a long term commitment to ensure effective control and to allow native vegetation to become established. Burma reed's deep roots make mechanical removal an extremely labor intensive and costly undertaking and causes extensive disturbance to the soil. A more effective management approach involves a combination of cutting or prescribed burning, followed by application of herbicides. After cutting, mowing or burning Burma reed plants down to the ground, a systemic herbicide like glyphosate, mixed with an acidic surfactant (trade name: ROUNDUP PRO) can be made to prevent new growth. Repeat treatment is likely to be necessary for a couple of years, until seed and rhizome stores are exhausted.

A successful burn of Burma reed reduces the plant's massive stalks to ash, eliminating the cost of vegetation removal. Conveniently, because Burma reed is the first plant to resprout following a fire, it can be sprayed freely with little concern about non-target kills. It should be noted that burning, by itself, whether through prescribed or natural wildfires, may enhance the growth and spread of Burma reed if not followed up with chemical or mechanical control.

In areas where Burma reed is dispersed among desirable native vegetation, individual plants can be cut at the base using a steel blade (e.g., Weed Wacker) and the remaining portions sprayed with ROUNDUP PRO to prevent new growth. Resprouts should be treated with a second herbicide application to the new growth. This method requires highly qualified applicators who can target the herbicide to avoid damage to native plants, and may not be cost effective for extensive infestations.

FOUNTAIN GRASS PENNISETUM SETACEUM (FORSK.) CHIOV

Native Range: Northern Africa

Description: Fountain grass is an attractive perennial grass with a densely clumped growth form and erect stems that grow 2 to 3 feet high. The small flowers of fountain grass are grouped in pink or purple, bristly, upright inflorescences 6-15 inches long. Fruits are small, dry achenes adorned with long showy bristles.

Ecological Threats Posed by Plant: Fountain grass is a highly aggressive, fire-adapted colonizer that readily outcompetes native plants and rapidly reestablishes after burning. In Hawaii, where it alters the natural fire regime, fountain grass is a major threat to some critically imperiled plant species and natural communities. Fountain grass raises fuel loads, which increases the intensity and spread of a fire, and results in severe damage to native, dry forest species adapted to less extreme fire regimes.

Distribution in the United States: Fountain grass is found in Arizona, California, Colorado, Hawaii, Florida, Louisiana, and Tennessee. In the Hawaiian islands, it is found on the islands of Kauai, Oahu, Lanai, and Hawaii.

Habitat in the United States: In Hawaii, fountain grass invades many types of natural areas, from bare lava flows to rangelands. It has a wide elevational range but is limited to areas with a median annual rainfall of less than 50 inches. In southern California, fountain grass invades grasslands, deserts, canyons and roadsides.

Background: First collected in Hawaii in 1914, fountain grass has been introduced to many parts of the world as an ornamental grass. It is a poor pasture grass and a serious weed in many dry habitats.

Biology and Spread: Fountain grass is a perennial plant with primarily wind-dispersed seeds that may remain viable in the soil for six years or longer. Its seeds may be dispersed greater distances by water, vehicles, livestock and humans.

Current Management Approaches: The long-lived seeds of fountain grass make its control extremely difficult. Small infestations may be managed by uprooting plants by hand and destroying the inflorescences in order to prevent seed dispersal. Removal by hand may need to be repeated several times per year. Extensive infestations of fountain grass are probably best controlled with the help of herbicides, especially those with some systemic activity.

JAPANESE KNOTWEED POLYGONUM CUSPIDATUM SIEB. & ZUCC.

Native Range: Eastern Asia

Description: Japanese knotweed, a member of the buckwheat family (Polygonaceae), is an upright, shrublike, herbaceous perennial that can grow to over 10 feet in height. As with all members of this family, the base of the stem above each joint is surrounded by a membranous sheath. Stems of Japanese knotweed are smooth, stout and swollen at joints where the leaf meets the stem. Although leaf size may vary, they are normally about 6 inches long by 3 to 4 inches wide, broadly oval to somewhat triangular and pointed at the tip. The minute greenish-white flowers occur in attractive, branched sprays in summer and are followed soon after by small winged fruits. Seeds arc triangular, shiny, and very small, about 1/10 inch long. Japanese knotweed is designated a noxious weed in most states.

Ecological Threat: Japanese knotweed spreads quickly to form dense thickets that exclude native vegetation and greatly alter natural ecosystems. It poses a significant threat to riparian areas, where it can survive severe floods and is able to rapidly colonize scoured shores and islands. Once established, populations are extremely persistent.

Distribution in the United States: Current distribution of Japanese knotweed includes 36 states in the lower 48, from Maine to Wisconsin south to Louisiana, and scattered Midwest and western states. It is not currently known to occur in Hawaii.

Habitat in the United States: Japanese knotweed can tolerate a variety of adverse conditions including full shade, high temperatures, high salinity, and drought. It is found near water sources, such as along streams and rivers, in low-lying areas, waste places, utility rights-of-way, and around old homesites. It can quickly become an invasive pest in natural areas after escaping from cultivated gardens.

Background: Japanese knotweed was probably introduced to the U.S. in the late 1800's. Also known as crimson beauty, Mexican bamboo, Japanese fleece flower, or Reynoutria, it was first introduced as an ornamental and has also been used for erosion control and for landscape screening. It is now found throughout the eastern U.S., in several western states, and Alaska, which has few exotic invasive plants to date.

Biology and Spread: Japanese knotweed spreads primarily by vegetative means with the help of its long, stout rhizomes. It is often transported to new sites as a contaminant in fill dirt; seeds are sometimes distributed by water, and carried to a lesser extent by the wind. Escapees from neglected gardens, and discarded cuttings are common routes of dispersal from urban areas.

Current Management Approaches: Grubbing is effective for small initial populations or environmentally sensitive areas where herbicides cannot be used. Using a Pulaski or similar digging tool, remove the entire plant including all roots and runners. Juvenile plants can be hand pulled depending on soil conditions and root development. Any portions of the root system not removed will potentially re-sprout. All plant parts (including mature fruit) should be bagged and disposed of in a trash dumpster to prevent re-establishment.

Cut stem treatment: Use this method in areas where plants are established within or around non-target plants or where vines have grown into the canopy. This treatment remains effective at low temperatures as long as the ground is not frozen. Cut the stem about 2 inches above ground level. Immediately apply a 25% solution of glyphosate (e.g., Roundup, or use Rodeo if applying in or near wetland areas) or triclopyr (e.g., Garlon) and water to the cross-section of the stem. A subsequent foliar application of glyphosate may be required to control new seedlings and resprouts. Foliar spray method: Use this method to control large populations. It may be necessary to precede foliar applications with stump treatments to reduce the risk of damaging non-target species. Apply a 2% solution of glyphosate or triclopyr and water to thoroughly wet all foliage. Do not apply so heavily that herbicide will drip off leaves. A 0.5% non-ionic surfactant is recommended in order to penetrate the leaf cuticle, and ambient air temperature should be above 65 °F.

LESSER CELANDINE RANUNCULUS FICARIA L.

Native Range: Europe

Description: Lesser celandine, also known as fig buttercup, is an herbaceous, perennial plant in the buttercup family (Ranunculaceae). Plants have a basal rosette of dark green, shiny, stalked leaves that are kidney-to heart-shaped. The flowers open in March and April, have eight glossy, butter-yellow petals, and are borne singly on delicate stalks that rise above the leaves. Pale-colored bulblets are produced along the stems of the above-ground portions of the plant, but are not apparent until late in the flowering period. When in bloom, large infestations of lesser celandine appear as a green carpet with yellow dots, spreading across the forest floor. There are many varieties of lesser celandine including a double-flowered form with many crowded petals and dark green leaves mottled with silvery markings.

NOTE: Lesser celandine closely resembles marsh marigold (Caltha palustris), a native wetland plant that occurs in the eastern United States. Marsh marigold is a robust plant with glossy, rounded or kidney-shaped leaves and flowers on stalks that are 8 in (20.3 cm) or more in height and consist of five to nine deep yellow "petals" (actually sepals). Marsh marigold does not produce tubers or bulblets, nor does it form a continuous carpet of growth. Extreme care should be taken to correctly identify lesser celandine before undertaking any control measures to avoid impacts to this plant.

Ecological Threat: Lesser celandine is an exotic spring ephemeral and a vigorous growing groundcover that forms large, dense patches on the forest floor, displacing and preventing native plants from co-occurring. The ecological impact of lesser celandine is primarily on the native spring-flowering plant community and the various wildlife species associated with them. Spring ephemerals complete the reproductive part of their life cycle and most of their above-ground development before woody plants leaf out and shade the forest floor. Native spring ephemerals include bloodroot, common and cut-leaved toothwort, Dutchman's breeches, harbinger-of-spring, squirrel-corn, trout lily, Virginia bluebells, and many others. Because lesser celandine emerges well in advance of the native species, it can establish and overtake areas rapidly.

Distribution in the United States: Lesser celandine is currently found in nineteen states in the Northeast and Pacific Northwest. It is reported to be invasive in nine states (Connecticut, Delaware, Maryland, New Jersey, Oregon, Pennsylvania, Virginia, Wisconsin, West Virginia), and the District of Columbia.

Habitat in the United States: Lesser celandine occurs in moist forested floodplains and in some drier upland areas, and seems to prefer sandy soils. **Background:** Lesser celandine was introduced to the United States as an ornamental plant. It is still available commercially in the U.S., along with many colorful varieties. All varieties of lesser celandine should be assumed to be invasive.

Biology and Spread: Lesser celandine is an exotic perennial plant and spring ephemeral that spends much of the year (summer through early winter) underground as thickened, fingerlike tubers or underground stems. During the winter, leaves begin to emerge and photosynthesize in preparation for flowering. Flowering usually occurs from late winter through mid-spring (March through May), depending on conditions. Afterwards, the above-ground portions die back. Lesser celandine spreads primarily by vegetative means through abundant tubers and bulblets, each of which is ready to become a new plant once separated from the parent plant. The tubers of lesser celandine are prolific and may be unearthed and scattered by the digging activities of some animals, including well-meaning weed pullers, and transported during flood events.

Current Management Approaches: Lesser celandine is very difficult to control but it can be managed with persistence over time using methods that are site appropriate. While manual methods are possible for some (small) infestations, the use of systemic herbicide kills the entire plant tip to root and minimizes soil disturbance. Biological. No biological control agents are currently available for lesser celandine. Chemical. The window of opportunity for controlling lesser celandine is very short, due to its life cycle. In order to have the greatest negative impact to celandine and the least impact to desirable native wildflower species, herbicide should be applied in late winter-early spring (March through May). Apply a 1.5% rate of a 39 to 41% glyphosate isopropylamine salt (e.g., Rodeo for wetland areas) mixed with water and a non-ionic surfactant to foliage, avoiding application to anything but the celandine. Glyphosate is systemic; that is, the active ingredient is absorbed by the plant and translocated to the roots, eventually killing the entire plant. The full effect on the plant may take 1-2 weeks. Applications can be made during the winter season as long as the temperature is above about 50°F, and no rain is anticipated within 12 hours. Because glyphosate is non-specific, spray should be controlled such that it touches only lesser celandine and does not drift onto desirable plants. To minimize impacts to sensitive-skinned frogs and salamanders, some experts recommend applying herbicide in March and then switching to manual methods. Manual. For small infestations, lesser celandine may be pulled up by hand or dug up using a hand trowel or shovel. It is very important to remove all bulblets and tubers. Mechanical. If mechanical removal is to continue after dieback of the plants, individual plants or clumps will need to be marked with some sort of stakes or flagging because it will be impossible to relocate the plants otherwise. When conducting mechanical removal, care should be taken to minimize soil disturbance as much as possible. For this reason, mechanical control may be inappropriate for large infestations in high quality natural areas.

COMMON MULLEIN VERBASCUM THAPSUS L.

Native Range: Europe and Asia

Description: Common mullein, also known as wooly mullein, is an erect herb in the figwort family, or Scrophulariaceae. First year mullein plants are low-growing rosettes of bluish gray-green, felt-like leaves that range from 4-12 inches in length and 1-5 inches in width. Mature flowering plants are produced the second year, and grow to 5 to 10 feet in height, including the conspicuous flowering stalk. The five-petaled yellow flowers are arranged in a leafy spike and bloom a few at a time from June-August. Leaves alternate along the flowering stalks and are much larger toward the base of the plant. The tiny seeds are pitted and rough with wavy ridges and deep grooves and can germinate after lying dormant in the soil for several decades.

Ecological Threat: Common mullein threatens natural meadows and forest openings, where it adapts easily to a wide variety of site conditions. Once established, it grows more vigorously than many native herbs and shrubs, and its growth can overtake a site in fairly short order. Common mullein is a prolific seeder and its seeds last a very long time in the soil. An established population of common mullein can be extremely difficult to eradicate.

Distribution in the United States: Common mullein was first introduced into the U.S. in the mid-1700's, where it was used as a piscicide, or fish poison, in Virginia. It quickly spread throughout the U.S. and is well established throughout the eastern states. Records show that it was first described in Michigan in 1839 and on the Pacific coast in 1876, probably due to multiple introductions as a medicinal herb.

Habitat in the United States: Common mullein can be found where mean annual precipitation is greater than 3-6 inches and the growing season lasts for a minimum of 140 days. Intolerant of shade, mullein will grow in almost any open area including natural meadows and forest openings as well as neglected pastures, road cuts, and industrial areas. Common mullein prefers, but is not limited to, dry sandy soils.

Background: Common mullein is a monocarpic perennial (i.e., takes two or more years to flower and die). Brought over from Europe by settlers, it was used as a medicinal herb, as a remedy for coughs and diarrhea and a respiratory stimulant for the lungs when smoked. A methanol extract from common mullein has been used as an insecticide for mosquito larvae.

Biology and Spread: During the first summer after germination mullein produces a tap root and a rosette of leaves. During this vegetative stage, the rosette increases in size during the growing season until low temperatures arrest growth sometime during the autumn and winter. Beginning the next spring, second year plants bolt into maturity, flower, produce seed during the summer, and then die, completing the plant's normal life cycle. Flowers mature from the base to the tip of the stalk. The length of the flowering period is a function of stalk height; longer stalks can continue to flower into early October.

It is estimated that a single plant can produce 100,000-180,000 seeds which may remain viable for more than 100 years. The seeds are dispersed mechanically near the parent plant during the autumn and winter. Seeds at or near the surface are more likely to germinate.

Current Management Approaches: Although common mullein can be very difficult to eradicate, there are a variety of management methods available, depending on the particular situation. Because mullein seedling emergence is dependent on the presence of bare ground, sowing sites with early successional native grasses or other plants may decrease seed germination and the chance of successful emergence of mullein seedlings. Mullein plants are easily hand pulled on loose soils due to relatively shallow tap roots. This is an extremely effective method of reducing populations and seed productivity, especially if plant is pulled before seed set. If blooms or seed capsules are present, reproductive structures should be removed, bagged, and properly disposed of in a sanitary landfill. Care should be taken, however, to minimize soil disturbance since loose soil will facilitate mullein seed germination.

There are two insects that have possible biological control implications for mullein. A European curculionid weevil (Gymnaetron tetrum), determined by the U.S. Department of Agriculture to be specific to mullein, has been introduced to North America. The weevil larvae matures in the seed capsules and can destroy up to 50% of the seeds. Another agent, the mullein moth (Cucullia verbasci) has been tested in the U.S. and is considered to be a relatively safe control agent because of its consistent feeding and development on mullein species. Although tests showed limited feeding on other native species, the larvae did not survive significantly longer than those individuals tested in the absence of food.

Release of biological controls into natural environments is always experimental and should be entered into only after full and careful consideration of potential non-target species impacts. Once released into nature, biological control agents are difficult, if not impossible, to control. For situations where hand-pulling of plants is not practical or safe--for example, on very steep slopes where hand pulling is dangerous or would cause significant soil disturbance--herbicidal control is an effective option. Apply a 2% solution of glyphosate (e.g., Roundup) or triclopyr (Garlon) and water plus a non-ionic surfactant, using a tank or backpack sprayer to thoroughly cover all leaves. Do not apply so heavily that the herbicide drips off the leaf surface. Use caution as glyphosate is a non-selective herbicide that may kill desirable plants even if partially contacted by spray. Triclopyr is selective to broadleaf plants and is a better choice if native or other desirable grasses are present. For some sites, applications can be made during the early spring when most other non-target vegetation is dormant. Refer to the pesticide manufacturers' label for specific information and restrictions regarding herbicide use.

VINE SECTION ALIEN PLANT INVADERS

Fiveleaf Akebia Akebia quinata (Houtt.) Dcne.

Native Range: Central China to Korea and Japan.

Description: Akebia quinata, also known as chocolate vine, is a twining woody vine in the mostly tropical Lardizabalaceae family. It grows as either a twining vine or vigorous groundcover and has slender, rounded stems that are green when young and brown at maturity. The palmate (like a hand) leaves of akebia alternate along the stem and are divided into five, or sometimes fewer, approximately equal parts called leaflets, whose small stems meet at a central juncture. Leaflets are generally long oval in shape, 1½ to 3 inches long and emerge with a purplish tinge, becoming blue-green at maturity. Flowers are unusual, chocolate-purple colored, and fragrant, about 1 inch across, and appear in late March to early April; they are often concealed by new foliage. Fruits are purpleviolet, flattened, sausage-like pods, 2 1/4 to 4 inches in length, that ripen in late September to early October. The inside of the pod has a whitish pulpy core with many tiny black seeds. Fiveleaf akebia is deciduous in cooler climates but may remain evergreen in the warmer regions, such as Louisiana.

Ecological Threat: Fiveleaf akebia grows so quickly that, if left unmanaged, it can kill off existing ground level vegetation, understory shrubs and trees, and even some canopy trees, by overtopping and smothering them. Once established, its dense growth prevents seed germination and seedling establishment of native plants.

Distribution in the United States: Fiveleaf akebia is found in 16 states in the eastern U.S., from Michigan to Connecticut, south to Georgia. Although it is not listed as occurring in South Carolina, this is probably an oversight.

Habitat in the United States: Akebia is shade and drought tolerant and can invade many types of habitats. Its growth appears to be restricted only by the height of the object it is entangling.

Background: Fiveleaf akebia brought to the United States in 1845 as an ornamental and has since naturalized in the warmer climates.

Biology and Spread: Akebia spreads primarily by vegetative means and is capable of growing twenty to forty feet in a single growing season. Fruits are not always produced and the seeds of akebia are not known to be carried by wind or insects. While birds may play a role in seed dispersal, fiveleaf akebia is vectored primarily through inadvertent activities of humans.

Current Management Approaches: Small infestations of fiveleaf akebia can be controlled mechanically. Because of its extremely rapid growth, cutting will need to be repeated throughout the season. At a minimum, akebia should be cut back to the ground at the end of the summer. Akebia vines may also be dug up, removing as much of the roots as possible. To ensure its complete removal, regular monitoring and repeated cutting, digging or pulling is necessary.

For large infestations, use of a labeled systemic herbicide, such as glyphosate (e.g., Roundup) or triclopyr (e.g., Garlon), is probably the most effective method to control akebia. An herbicidal soap, such as pelargonic acid (e.g., Scythe), which provides a burndown of plant tissues, may also provide some control.

The latter product may also provide enhanced uptake of certain systemic herbicides, such as glyphosate. Be sure to avoid drift onto or direct spray of desirable plant species during herbicide applications.

Suggested Alternative Plants: Some native alternative vines for the eastern U.S. include trumpet honeysuckle (Lonicera sempervirens), cross vine (Bignonia capreolata), trumpet creeper (Campsis radicans), or Dutchman's pipe (Aristolochia durior). Contact your local native plant society for other suggestions.

PORCELAINBERRY AMPELOPSIS BREVIPEDUNCULATA (MAXIM.) TRAUTV.

Native Range: China, Korea, Japan, Russian Far East

Description: Porcelainberry is a deciduous, woody, perennial vine of the grape family, Vitaceae. It is related to the North American raccoon-grape and peppervine, and is sometimes referred to as Amur peppervine or porcelain ampelopsis. The simple, heart-shaped leaves of porcelainberry are dark green with coarsely toothed edges, and are shiny underneath with delicate hairs along the veins. The leaves vary from slightly 3-5 lobed to deeply dissected, the latter being distinctively recognizable. The leaves are arranged alternately on vines that grow to heights of 15 to 20 feet. The plant climbs by tendrils that grow opposite the leaves on the stem. Small, greenish-white flowers appear in clusters on porcelainberry in June, July, and August. The colorful berries of porcelainberry, its most distinguishing feature, appear in September-October. The berries, about 1/4-inch in diameter, range in color from white to yellow, to pastel shades of green, lilac and amethyst purple, to turquoise and sky blue. All colors of the berry are often found growing on the same plant at the same time while the plant is still in full foliage. There are several US native species of Ampelopsis that could be confused with this exotic.

Ecological Threat: Porcelainberry is a vigorous vine which is slow to establish but grows and spreads quickly in open areas of the urban landscape. The seeds of porcelainberry are similar in size to those of native pioneers, and like the native, they will germinate readily in the prepared soil bed left behind after natural or human disturbance. Once established, the vine quickly overwhelms and destroys native vegetation by shading out smaller plants and outcompeting native vegetation for water and nutrients. Urban parks, with extensive wooded borders neighboring landscaped residential and private property, are especially vulnerable to invasion by porcelainberry.

Distribution in the United States: New England to North Carolina and west to Michigan.

Habitat in the United States: Porcelainberry grows well in most soils, especially in pond margins, streambanks, thickets, and waste places, where there is full sunlight to partial shade, and where it is not permanently wet. Like its relative the grape, porcelainberry appears to be less tolerant of heavily shaded areas, such as that found in mature forest.

Background: Porcelainberry was originally cultivated as a bedding and landscape plant. In spite of its aggressiveness in some areas, it is still used in the horticultural trade (for example, the ornamental A. brevipedunculata 'Elegans' is often recommended as a landscape plant with a cautionary note that "care must be taken to keep it from overtaking and shading out small plants"). The characteristics that make porcelainberry a desirable plant for the garden--its colorful berries, good ground coverage, trellis-climbing vines, pest-resistance, and tolerance of adverse conditions--are also responsible for its presence in the United States as an undesirable invader.

Biology and Spread: The colorful berries (carrying 2-4 seeds each) attract birds and other small animals that eat the berries and disperse the seeds in their droppings. Because the plant is often found growing in riparian areas downstream from established plant colonies, it is thought that the seeds may also be dispersed in stream water.

Current Management Approaches: Since the vines can grow as long as 15 ft. in a single growing season, repeated applications in the same growing season and in subsequent years may be necessary to fully eradicate the plant.

Manual. Because flowers are produced on the current season's growth, hand-pruning in the fall or spring will prevent flower buds from forming the following season. Where feasible, plants should be pulled up by hand before fruiting to prevent the spread of seeds. If the plants are pulled while in fruit, the fruits should be bagged and burned before they ripen. This method can be difficult in areas where the plant is well established since pulling up the plant often pulls out native plants with intermingled roots. Cut aboveground vines and, if possible, pull from trees to allow the trees to recover and re-cut repeatedly as needed.

Chemical. The herbicides triclopyr (e.g., Garlon 3a and Garlon 4) and glyphosate (Roundup and Rodeo) have been used with varying success to battle infestations of porcelainberry. Foliar applications: Smaller infestations of porcelainberry can be controlled to some extent with spot applications of glyphosphate to leaves, used sparingly to avoid contact of desirable plants with spray. Cut vines back during the summer and allow to resprout before applying herbicide, or apply glyphosate to leaves in early autumn, just prior to senescence. More effective control has been achieved using triclopyr formulations. From summer to fall, apply a water-based solution of 2.5% Garlon 3A (triclopyr amine) to foliage or cut first, allow to re-grow, and then apply triclopyr to new growth.

Basal bark applications: Apply a mixture of 20-30% Garlon 4 (triclopyr ester) with an equal volume of commercially available basal oil, diesel fuel, No. 1 or 2 fuel oil, or kerosene. Other oils, such as horticultural oil, can be substituted if the label for the oil recommends it for basal bark application. Application should be during any season when temperatures are around 60°F or more for several days; apply around the basal portions of vines.

ORIENTAL BITTERSWEET CELASTRUS ORBICULATUS THUNB.

Native Range: Eastern Asia, Korea, China and Japan

Description: Oriental bittersweet is a deciduous, woody, perennial vine in the stafftree family (Celastraceae), which sometimes occurs as a trailing shrub. Also known as round-leaved and Asiatic bittersweet, stems of older plants sometimes grow to four inches in diameter. Leaves of oriental bittersweet are glossy, rounded, finely toothed and arranged alternately along the stem. Clusters of small greenish flowers emerge from leaf axils, allowing each plant to produce large numbers of seeds. At maturity, globular, green to yellow fruits split open to reveal three red-orange, fleshy arils that contain the seeds. These showy fruits have made oriental bittersweet very popular for use in floral arrangements. Since this plant is easily confused with our native climbing bittersweet vine (Celastrus scandens), which flowers at the tips rather than along the stems, it is imperative that correct identification be made before controls are attempted.

Ecological Threat: Oriental bittersweet is an aggressive invader that threatens all vegetation levels of forested and open areas. It grows over other vegetation, completely covering it, and kills other plants by preventing photosynthesis, girdling, and uprooting by force of its massive weight. In the northeastern U.S., exotic Oriental bittersweet appears to be displacing the native climbing bittersweet (Celastrus scandens), which occurs in similar habitats, through competition and hybridization.

Distribution in the United States: Oriental bittersweet currently occurs from New York to North Carolina, and westward to Illinois.

Habitat in the United States: Oriental bittersweet infests forest edges, woodlands, early successional fields, hedgerows, coastal areas and salt marsh edges, particularly those suffering some form of land disturbance. While often found in more open, sunny sites, its tolerance for shade allows oriental bittersweet to invade forested areas.

Background: Introduced into the U.S. in the 1860s as an ornamental plant, oriental bittersweet is often associated with old homesites, from which it has escaped into surrounding natural areas. Oriental bittersweet is still widely planted and maintained as an ornamental vine, further promoting its spread.

Biology and Spread: Oriental bittersweet reproduces prolifically by seed, which is readily dispersed to new areas by many species of birds. Its seeds germinate in late spring in partial to dense shade. It also expands vegetatively by stolons (above-ground stems), and rhizomes (underground stems), and through root suckering (the ability to send shoots up from the roots).

Current Management Approaches: Where hand labor is practical, vines can be pulled out by the roots and removed from the site, preferably before fruiting. If fruits are present, vines should be bagged and disposed of in a landfill, or left in the bags and allowed to bake in the sun long enough to kill the seeds. Certain systemic herbicides, such as glyphosate (e.g., Roundup) or triclopyr (e.g., Garlon), that are taken into the roots and kill the entire plant, have been used successfully in bittersweet management. This method is most effective if the stems are first cut by hand or mowed and herbicide is applied immediately to cut stem tissue. In areas where spring wildflowers or other native plants occur, application of herbicides should be conducted prior to their emergence, delayed until late summer or autumn, after the last killing frost occurs, or carefully targeted. Herbicidal contact with desirable plants should always be avoided. No biological controls are currently known for oriental bittersweet.

Suggested Alternative Plants: Several attractive native vines are available that provide nectar, seed and host plant material for butterflies, hummingbirds, and other wildlife. These include common, or climbing, bittersweet (Celastrus scandens) which is native to the eastern U.S., trumpet creeper (Campsis radicans), passionflower vine (Passiflora lutea), pipevine or Dutchman's pipe (Aristolochia macrophylla) and native wisteria* (Wisteria frutescens). *If you wish to plant wisteria, make certain that it is the native species. Two commonly planted ornamental wisterias, Chinese wisteria (Wisteria sinensis) and Japanese wisteria (Wisteria floribunda), are exotic and aggressive invaders.

CLIMBING EUONYMUS EUONYMUS FORTUNEI (TURCS.) HAND.-MAZZ.

Native Range: China

Description: Climbing euonymus, also known as wintercreeper, Emerald'n Gold, and Gaiety, is an evergreen, clinging vine in the staff-tree (Celastraceae) family. It can form a dense groundcover or shrub to 3 feet in height, or climb 40-70 foot high vertical surfaces with the aid of aerial roots. Dark green, shiny, egg-shaped leaves, from 1-2 1/2 inches long, with toothed margins and silvery veins, occur in pairs along the stems. Stems are narrow, minutely warty, and have abundant rootlets or trailing roots. Clusters of inconspicuous green-white flowers are produced on a long stalk from June to July and are followed in the autumn by pinkish to red capsules that split open to expose seeds adorned with a fleshy orange seed coat, or aril.

Ecological Threat: Traits that make climbing euonymus a desirable ornamental plant, such as its rapid growth, evergreen nature and tolerance of harsh conditions, also make euonymus a threat to natural areas. Climbing euonymus can outcompete native vegetation by depleting soil moisture and nutrients, blocking sunlight, and by forming a dense vegetative mat that impedes the growth of seedlings of native species. Vines on trees continue climbing and can eventually overtop them, covering the leaves and preventing photosynthesis.

Distribution in the United States: Climbing euonymus is currently scattered throughout the eastern U.S. in populated areas.

Habitat in the United States: Climbing euonymus tolerates a variety of environmental conditions, including poor soils, full sun to dense shade, and a wide pH range. It does not do well in heavy, wet soils. Natural forest openings resulting from wind throw, insect defoliation or fire are vulnerable to invasion and provide conditions for satellite populations of climbing euonymus to get started.

Background: Climbing euonymus was introduced into the U.S. in 1907 as an ornamental ground cover.

Biology and Spread: Climbing euonymus spreads vegetatively with the help of lateral shoots produced along its long main branches and by new plants that emerge from rootlets also produced along the stem at short intervals. Vines climb rocks, trees, and other supporting structures. Flowers formed in the summer produce mature fruits by fall that are equipped with fleshy edible structures (arils) that are fed on by birds and other wildlife which disperse it. Climbing euonymus also escapes from neglected gardens and is carried by water to undisturbed forest and riparian areas.

Current Management Approaches: A variety of mechanical and chemical methods are available for management of climbing euonymus. Grubbing, a rather labor intensive method, is effective for small populations or environmentally sensitive areas where herbicides cannot be used. Using a Pulaski or similar digging tool, remove the entire plant, including all roots and runners. Juvenile plants can be hand-pulled when the soil is moist and root systems are small. Any portions of the root system remaining may resprout. All plant parts including stem fragments and mature fruits should be bagged and disposed of in a trash dumpster to prevent reestablishment.

Cut stem treatment, using systemic herbicides applied to freshly cut stems, is effective in areas where vines are well established on or around non-target plants, or where they have grown into tree canopies or other vertical surfaces. Cut the stem as close to the ground as possible and immediately apply a 25% solution of glyphosate (e.g., Roundup) or triclopyr (e.g., Garlon) and water to the cut stem. This procedure is effective at temperatures as low as 40°F. Subsequent foliar application of these herbicides may be required. Cutting without the application of herbicides is generally not recommended because this will lead to root sprouting. Foliar applications of herbicide can be used to control large populations.

It may be necessary to precede foliar sprays with cut stem treatments to reduce the risk of damage to non-target plants. Apply a 2% solution of glyphosate or triclopyr and water plus a 0.5% non-ionic surfactant to thoroughly wet all foliage but not so heavily that it drips off leaves where it may affect desirable plants. Glyphosate is a non-selective systemic (i.e., travels through the plant vessels) herbicide that may kill even partially sprayed plants. Triclopyr is selective to broad leaf species and is a better choice if desirable native grasses are present. Ambient air temperature should be above 65°F.

Suggested Alternative Plants: There are a variety of native creeping or climbing vines that make good alternatives for climbing euonymus. Some examples from the eastern U.S. include trumpet creeper (Campsis radicans), Dutchman's pipe (Aristolochia macrophylla), crossvine (Bignonia capreolata), trumpet honeysuckle (Lonicera sempervirens), American bittersweet (Celastrus scandens), and American wisteria (Wisteria frutescens), our only native wisteria*.

ENGLISH IVY HEDERA HELIX L.

Native Range: Europe, western Asia, and northern Africa

Description: English ivy is an evergreen climbing vine in the ginseng family (Araliaceae). Vines attach to the bark of trees, brickwork, and other surfaces by way of numerous, small rootlike structures, which exude a glue-like substance. Older vines are known to reach a foot in diameter. Leaves are dark green, waxy, somewhat leathery, and are arranged alternately along the stem. English ivy has many recognized leaf forms, the most common being a 3-lobed leaf with a heartshaped base. Leaves in full sun are often unlobed, oval and have wedge-shaped bases. Umbrella-like clusters of small, greenish-white flowers appear in the fall if sufficient sunlight is available. Fruits mature in Spring and are black with a fleshy outer covering enclosing one to a few hard, stone-like seeds.

Note: Compounds in English ivy are somewhat toxic and include glycosides that cause vomiting, diarrhea, nervous conditions and dermatitis in sensitive individuals. This characteristic helps ensure spread of the seeds by many native songbirds that are attracted to the black berries in Spring when other food sources are limited.

Ecological Threat: English ivy is an aggressive invader that threatens all vegetation levels of forested and open areas, growing along the ground as well as into the forest canopy. The dense growth and abundant leaves, which spring from the stems like small umbrellas, form a thick canopy just above the ground, and prevent sunlight from reaching other plants. Similarly, vines climbing up tree trunks spread out and surround branches and twigs, preventing most of the sunlight from reaching the leaves of the host tree. Loss of host tree vigor, evident within a few years, is followed by death a few years later.

The added weight of vines makes infested trees susceptible to blow-over during storms. English ivy also serves as a reservoir for bacterial leaf scorch (Xylella fastidiosa), a plant pathogen that is harmful to native trees such as elms, oaks, and maples. English ivy is a popular plant, recommended by Cooperative Extension offices for use as a low maintenance alternative to lawns. It is widely used by homeowners, horticulturists, landscape contractors, parks departments and others desiring a fast-growing, low maintenance, evergreen groundcover. Once established at a site, English ivy can be expected to move beyond its intended borders into neighboring yards, parks and other lands, either by vegetative means or by seed.

Distribution in the United States: English ivy occurs in at least 26 states and the District of Columbia, where it is one of the most abundant and widespread invasive plants.

Habitat in the United States: English ivy infests woodlands, forest edges, fields, hedgerows, coastal areas, salt marsh edges, and other upland areas, especially where some soil moisture is present. It does not grow well in extremely wet conditions and is often associated with some form of land disturbance, either human-caused or natural.

Background: English ivy was probably first introduced to the US by European immigrants and is widely sold as an ornamental plant for landscapes throughout the US.

Biology and Spread: English ivy reproduces vegetatively and by seed, which is dispersed to new areas primarily by birds, including English house sparrows, European starlings, robins, Stellar jays, and cedar waxwings. New plants grow easily from cuttings or from stems making contact with the soil.

Current Management Approaches: Several effective methods of control are available for English ivy, including chemical and non-chemical, depending on the extent of the infestation, the amount of native vegetation on-site, and available time and labor. Manual and Mechanical. Vines growing as groundcover can be pulled up by hand, with some difficulty, and left on-site or bagged and disposed of as trash. Vines climbing up into the tree canopy are more difficult to manage. First, vines should be cut at a comfortable height to kill upper portions and relieve the tree canopy.

A large screw driver or forked garden tool can be used to pry and snap the vines away from the tree trunks. Vines can be cut using an axe or, with more difficulty, using a pruning saw. Rooted portions of vines will remain alive and should be pulled and repeatedly cut. Because cutting will likely promote further growth from the base, vigilance is required to ensure long term control.

Chemical: The systemic herbicide triclopyr (e.g., Garlon) is absorbed into plant tissues and carried to the roots, effectively killing the entire plant in place.

Foliar Applications: From summer to fall, apply a 2.5% mixture of triclopyr amine (Garlon 3A) in water to the leaves or cut first, allow to re-grow, and apply the same mix to new foliage. Herbicide will also be absorbed through the stem bark for additional effect.

Basal Bark Applications: A higher rate (15-30%) of triclopyr ester (Garlon 4) may also be applied to stems of vines growing up trees but there is a possibility that the herbicide will be absorbed into the host tree, depending on the thickness of the host tree's bark and the penetration of English ivy rootlets.

Because English ivy is an evergreen vine, and remains active during the winter, herbicide applications can be made to it any time of year as long as temperatures are above 55 or 60°Fahrenheit for a few days. Fall and winter applications will avoid or minimize impacts to many native plant species. Repeat herbicidal treatments are likely to be needed and follow-up monitoring should be conducted to evaluate the success of treatments. Herbicidal contact with desirable plants should always be avoided. In areas where spring wildflowers or other native plants are interspersed, application of herbicides should be conducted prior to their emergence, or delayed until they have died back Biological control. There are no biological controls currently available for English ivy.

JAPANESE HONEYSUCKLE LONICERA JAPONICA THUNB.

Native Range: Japan and Korea

Description: Japanese honeysuckle, a member of the honeysuckle family (Caprifoliaceae), is a perennial vine that climbs by twisting its stems around vertical structures, including limbs and trunks of shrubs and small trees. Leaves are oblong to oval, sometimes lobed, have short stalks, and occur in pairs along the stem. In southern and mid-Atlantic states, Japanese honeysuckle often remains evergreen-its leaves remain attached through the winter. In colder northern climates, the leaves may fall off after exposure to prolonged winter temperatures. Flowers are tubular, with five fused petals, white to pink, turning yellow with age; they are very fragrant, and occur in pairs along the stem at leaf junctures. Stems and leaves are sometimes covered with fine, soft hairs. Japanese honeysuckle blooms from late April through July and sometimes into October. Small black fruits are produced in autumn, each containing 2-3 oval to oblong, dark brown seeds about 1/4 inch across.

Ecological Threat: In North America, Japanese honeysuckle has few natural enemies which allows it to spread widely and out-compete native plant species. Its evergreen to semi-evergreen nature gives it an added advantage over native species in many areas. Shrubs and young trees can be killed by girdling when vines twist tightly around stems and trunks, cutting off the flow of water through the plant. Dense growths of honeysuckle covering vegetation can gradually kill plants by blocking sunlight from reaching their leaves. Vigorous root competition also helps Japanese honeysuckle spread and displace neighboring native vegetation.

Distribution in the United States: Japanese honeysuckle occurs across the southern U.S. from California to New England and the Great Lakes region. Escaped populations also occur in Hawaii. Severe winter temperatures and low precipitation may limit its distribution in northern latitudes and in the West, respectively.

Habitat in the United States: A ubiquitous invader, Japanese honeysuckle thrives in a wide variety of habitats including fields, forests, wetlands, barrens, and all types of disturbed lands.

Background: Japanese honeysuckle was introduced to the U.S. in the early to mid-1800's as an ornamental plant, for erosion control, and for wildlife forage and cover. Its highly fragrant flowers provide a tiny drop of honey-flavored nectar enjoyed by children.

Biology and Spread: Growth and spread of Japanese honeysuckle is through vegetative (plant growth) and sexual (seed) means. It produces long vegetative runners that develop roots where stem and leaf junctions (nodes) come in contact with moist soil. Underground stems (rhizomes) help to establish and spread the plant locally. Long distance dispersal is by birds and other wildlife that readily consume the fruits and defecate the seeds at various distances from the parent plant.

Current Management Approaches: Several effective methods of control are available for Japanese honeysuckle, including chemical and non-chemical, depending on the extent of the infestation and available time and labor.

Manual and Mechanical: For small patches, repeated pulling of entire vines and root systems may be effective. Hand pull seedlings and young plants when the soil is moist, holding low on the stem to remove the whole plant along with its roots.

Monitor frequently and remove any new plants. Cut and remove twining vines to prevent them from girdling and killing shrubs and other plants. An effective method for removal of patches of honeysuckle covering the ground is to lift up and hold a portion of the vine mass with a rake and have a chain saw operator cut the stems low to the ground. Mowing large patches of honeysuckle may be useful if repeated regularly, but is most effective when combined with herbicide application (see below). Mow at least twice a year, first in mid-July and again in mid-September. Plants can also be grubbed out using a Pulaski or similar digging tool, taking care to remove all roots and runners. Burning removes above ground vegetation but does not kill the underground rhizomes, which will continue to sprout. In certain situations, tethered goats have been used to remove honeysuckle growth, but must be monitored to prevent their escape to the wild where they would become an added ecological threat. Chemical. In moderate cold climates, Japanese honeysuckle leaves continue to photosynthesize long after most other plants have lost their leaves. This allows for application of herbicides when many native species are dormant. However, for effective control with herbicides, healthy green leaves must be present at application time and temperatures must be sufficient for plant activity. Several systemic herbicides (e.g., glyphosate and triclopyr) move through the plant to the roots when applied to the leaves or stems and have been used effectively on Japanese honeysuckle.

Following label guidelines, apply a 2.5% rate of glyphosate (e.g., Rodeo for wetlands; Roundup for uplands) mixed with water and an appropriate surfactant, to foliage from spring through fall. Alternatively, apply a 2% concentration of triclopyr (e.g., Garlon 3A) plus water to foliage, thoroughly wetting the leaves but not to the point of drip-off. A coarse, low-pressure spray should be used.

Repeat applications may be needed. Treatment in the fall, when many non-target plants are going dormant, is best. Also, a 25% glyphosate or triclopyr solution mixed with water can be applied to cut stem surfaces any time of year as long as the ground is not frozen. Biological control. No biological control agents are currently available for Japanese honeysuckle.

Suggested Alternative Plants: Vines that make good substitutes for Japanese honeysuckle include false jasmine (Gelsemium sempervirens), trumpet honeysuckle (Lonicera sempervirens), trumpet creeper (Campsis radicans), crossvine (Bignonia capreolata), native wisteria (Wisteria frutescens), jackman clematis (Clematis jackmanii), and others. Check with your state native plant society, a reputable native plant nursery, for recommendations for plants that are appropriate for your area and conditions.

MILE-A-MINUTE WEED POLYGONUM PERFOLIATUM L.

Native Range: India to Eastern Asia, China and the Islands from Japan to the Philippines, including Nepal, Burma, Manchuria, China, Korea, Taiwan and the Malay Peninsula

Description: Mile-a-minute weed, also known as Devil's tail tearthumb, is an herbaceous, annual, trailing vine in the buckwheat family, Polygonaceae. It has a reddish stem that is armed with downward pointing hooks or barbs which are also present on the underside of the leaf blades.

The light green colored leaves are shaped like an equilateral (equal-sided) triangle and alternate along the narrow, delicate stems. Distinctive circular, cupshaped leafy structures, called ocreas, surround the stem at intervals. Flower buds, and later flowers and fruits, emerge from within the ocreas. Flowers are small, white and generally inconspicuous. The fruits are attractive, metallic blue and segmented, each segment containing a single glossy, black or reddish-black seed.

Ecological Threat: Mile-a-minute weed grows rapidly, scrambling over shrubs and other vegetation, blocking the foliage of covered plants from available light, and reducing their ability to photosynthesize, which stresses and weakens them. If left unchecked, the lack of photosynthesis will kill a plant. Large infestations of mile-a-minute weed eventually reduce native plant species in natural areas. Small populations of extremely rare plants may be eliminated entirely. Because it can smother tree seedlings, mile-a-minute weed has a negative effect on Christmas tree farms, forestry operations on pine plantations and reforestation of natural areas. It has the potential to be a problem to nursery and horticulture crops that are not regularly tilled as a cultivation practice.

Distribution in the United States: Mile-a-minute weed is currently found in Pennsylvania, Maryland, Delaware, West Virginia, New York, Virginia, Ohio and Washington, D.C. These states and the District of Columbia comprise about 20 percent of the estimated possible range for this species. It is considered a temperate species with subtropical tendencies and therefore has the potential to invade those portions of the contiguous United States that have the appropriate climate to provide a minimal eight week cold vernalization period. A temperature of 10°C or below must be sustained for an eight week period to stimulate germination.

Habitat in the United States: Mile-a-minute weed generally colonizes open and disturbed areas, along the edges of woods, wetlands, stream banks, and roadsides, and uncultivated open fields, resulting from both natural and human causes. Natural areas such as stream banks, parks, open space, road shoulders, forest edges and fence lines are all typical areas to find mile-a-minute. It also occurs in environments that are extremely wet with poor soil structure. Available light and soil moisture are both integral to the successful colonization of this species. It will tolerate shade for part of the day, but needs a good percentage (63-100%) of available light. The ability of mile-a-minute to attach to other plants with its recurved barbs and climb over the plants to reach an area of high light intensity is a key to its survival. It can survive in areas with relatively low soil moisture, but demonstrates a preference for high soil moisture.

Background: The first records of mile-a-minute in North America are from Portland, Oregon (1890) and Beltsville, Maryland (1937). Both of these sites were eliminated or did not establish permanent populations of the species. However, the introduction of mile-a-minute in the late 1930's to a nursery site in York County, Pennsylvania did produce a successful population of this plant. It is speculated that the seed was spread with Rhododendron stock. The owner of the nursery was interested in the plant and allowed it to reproduce; unfortunately, subsequent efforts to eradicate it were not successful. The distribution of mile-a-minute has radiated from the York County site into neighboring states. In the past 55 years, the range for this plant in the United States has extended as far as 300 miles in several directions from the York County, Pennsylvania site (Mountain, 1995) and (Okay 1997).

Biology and Spread: Mile-a-minute weed is primarily a self-pollinating plant (supported by its inconspicuous, closed flowers and lack of a detectable scent), with occasional outcrossing. Fruits and viable seeds are produced without assistance from pollinators. Vegetative propagation from roots has not been successful for this plant. It is a very tender annual, withering with a slight frost, and reproduces successfully until the first frost. Mile-a-minute is a prolific seeder, producing many seeds on a single plant over a long season, from June until October in Virginia, and a slightly shorter season in more northern geographic areas.

Birds are probably the primary long-distance dispersal agents of mile-a-minute seeds. Transport of seeds short distances by native ant species has been observed. This activity is probably encouraged by the presence of a tiny white food body (elaiosome) on the tip of the seed that may be attractive to the ants. These seed-carrying ants may play an important role in the survival and germination of the seeds of mile-a-minute weed. Local bird populations are important for dispersal under utility lines, bird feeders, fence lines and other perching locations. Other animals observed eating mile-a-minute weed fruits are chipmunks, squirrel and deer. Water is also an important mode of dispersal for mile-a-minute weed. Its fruits can remain buoyant for 7-9 days, an important advantage for dispersing seed long distances in stream and river environments. The long vines frequently hang over waterways, allowing fruits that detach to be carried away in the water current. During storm events the potential spread of this plant is greatly increased throughout watersheds.

Current Management Approaches: A variety of control measures, including physical, mechanical, cultural and chemical, can be used for management of mile-a-minute weeds. Hand pulling of seedlings is best done before the recurved barbs on the stem and leaves harden. But may be done afterwards with the help of thick gloves. Removal of vines by hand may be conducted throughout the summer, if tough gloves and protective clothing (coveralls) are worn to avoid the skin shredding ability of the recurved hooks. The delicate vines can be reeled in fairly easily and balled up in piles that can be left to dehydrate for several days before disposal. The site must be rechecked at frequent intervals, and removal of new plants continued until the seed germination period is complete, roughly early April until early July in the middle Atlantic states. Repeated mowing or trimming of mile-a-minute plants will prevent the plants from flowering and thus reduce or eliminate fruit and seed production. Cultural methods can be utilized to discourage the introduction of mile-a-minute to an area. It is important to maintain vegetative community stability and to avoid creating gaps or openings in existing vegetation. Maintaining broad vegetative buffers along streams and forest edges will help to shade out and prevent establishment of mile-a-minute weed. This will also help to reduce the dispersal of fruits by water.

An herbicidal soap, of which there are several brands available at most garden centers, will help to burn back foliage of mile-a-minute. Because these products do not have the systemic (i.e., travels to the roots) ability of herbicides like glyphosate, they will need to be reapplied all season long to any regrowth. Glyphosate (e.g., Roundup for upland areas and Rodeo for wetland applications), applied at a low rate will probably be effective in killing mile-a-minute weed. However, because this plant is not currently listed on the product labels for Roundup or Rodeo, treatments with these products is permissible only with prior approval of the State Department of Agriculture where the application will take place [FIFRA 1997, Section 2(z)(ee)].

KUDZU PUERARIA MONTANA VAR. LOBATA (WILLD.) MAESEN & S. ALMEIDA

Native Range: Asia

Description: Kudzu is a climbing, semi-woody, perennial vine in the pea family. Deciduous leaves are alternate and compound, with three broad leaflets up to 4 inches across. Leaflets may be entire or deeply 2-3 lobed with hairy margins. Individual flowers, about 1/2 inch long, are purple, highly fragrant and borne in long hanging clusters. Flowering occurs in late summer and is soon followed by production of brown, hairy, flattened, seed pods, each of which contains three to ten hard seeds.

Ecological Threat: Kudzu kills or degrades other plants by smothering them under a solid blanket of leaves, by girdling woody stems and tree trunks, and by breaking branches or uprooting entire trees and shrubs through the sheer force of its weight. Once established, Kudzu plants grow rapidly, extending as much as 60 feet per season at a rate of about one foot per day. This vigorous vine may extend 32-100 feet in length, with stems 1/2-4 inches in diameter. Kudzu roots are fleshy, with massive tap roots 7 inches or more in diameter, 6 feet or more in length, and weighing as much as 400 pounds. As many as thirty vines may grow from a single root crown.

Distribution in the United States: Kudzu is common throughout most of the southeastern U.S. and has been found as far north as Pennsylvania.

Habitat in the United States: Kudzu grows well under a wide range of conditions and in most soil types. Preferred habitats are forest edges, abandoned fields, roadsides, and disturbed areas, where sunlight is abundant. Kudzu grows best where winters are mild, summer temperatures are above 80°F, and annual rainfall is 40 inches or more.

Biology and Spread: The spread of kudzu in the U.S. is currently limited to vegetative expansion by runners and rhizomes and by vines that root at the nodes to form new plants. Kudzu also spreads somewhat through seeds, which are contained in pods, and which mature in the fall. However, only one or two viable seeds are produced per cluster of pods and these hard-coated seeds may not germinate for several years.

Background: Kudzu was introduced into the U.S. in 1876 at the Philadelphia Centennial Exposition, where it was promoted as a forage crop and an ornamental plant. From 1935 to the mid-1950s, farmers in the south were encouraged to plant kudzu to reduce soil erosion, and Franklin D. Roosevelt's Civilian Conservation Corps planted it widely for many years. Kudzu was recognized as a pest weed by the U.S. Department of Agriculture and, in 1953, was removed from its list of permissible cover plants.

Current Management Approaches: For successful long term control of kudzu, the extensive root system must be destroyed. Any remaining root crowns can lead to reinfestation of an area. Mechanical methods involve cutting vines just above ground level and destroying all cut material. Close mowing every month for two growing seasons or repeated cultivation may be effective. Cut kudzu can be fed to livestock, burned or enclosed in plastic bags and sent to a landfill. If conducted in the spring, cutting must be repeated as regrowth appears to exhaust the plant's stored carbohydrate reserves. Late season cutting should be followed up with immediate application of a systemic herbicide (e.g., glyphosate) to cut stems, to encourage transport of the herbicide into the root system. Repeated applications of several soil-active herbicides have been used effectively on large infestations in forestry situations. Efforts are being organized by the U.S. Forest Service to begin a search for biological control agents for kudzu.

Suggested Alternative Plants: Native vines such as trumpet creeper (Campsis radicans), pipevine (Aristolochia macrophylla), passionflower (Passiflora lutea), trumpet honeysuckle (Lonicera sempervirens), and native bittersweet (Celastrus scandens) have attractive flowers and fruits, provide food for wildlife and make excellent substitutes for kudzu. These plants should be used in landscaping and for land restoration where they are known to occur as natives.

CHINESE WISTERIA WISTERIA SINENSIS (SIMS) DC.

Japanese Wisteria Wisteria floribunda (Willd.) DC.

Native Range: China and Japan

Description: Chinese and Japanese wisteria are exotic, showy, woody ornamental vines in the pea family, or Fabaceae.

These vigorous vines can climb trees, apparently limited only by the height of the tree, and have been observed to reach 65 feet. Unlike American wisteria (Wisteria frutescens), native to the southeastern U.S., which flowers June through August, and produces a non-hairy seed pod 2-4 inches long, both exotic wisterias flower in the springtime (April-May) and produce a velvety seed pod. The fuzzy brown seed pods are 4-6 inches long, narrowed toward the base, with constrictions between the seeds. Stems of the exotic wisterias can grow to 15 inches in diameter in older plants. White-barked Japanese wisteria vines twine clockwise around the host plant and Chinese wisteria twines counter-clockwise. The compound leaves, consisting of 7-13 (Chinese) or 13-19 (Japanese) smaller leaf units, called leaflets, are about 1 foot long and alternate along the stem. Fragrant, violet to blue-violet flowers, ½ to 1 inch long, occur in showy, pendulous clusters that hang gracefully from the twining stems.

Ecological Threat: Exotic wisterias impair and overtake native shrubs and trees through strangling or shading. Climbing wisteria vines can kill sizable trees, opening the forest canopy and increasing sunlight to the forest floor, which in turn favors its aggressive growth. Chinese and Japanese wisterias are hardy and aggressive, capable of forming thickets so dense that little else grows.

Distribution in the United States: Japanese and Chinese wisteria are found extensively throughout the eastern states.

Habitat in the United States: The ideal habitat for exotic wisterias is in full sun, but established vines will persist and reproduce in partial shade. Vines often climb surrounding vegetation and structures toward sunlight. Wisteria tolerates a variety of soil and moisture regimes but prefers loamy, deep, well drained soils. Infestations are commonly found along forest edges, roadsides, ditches, and rights-of-way.

Background: Chinese wisteria was brought to the U.S. from China in 1816 and Japanese wisteria was introduced from Japan around 1830. Both were brought in as ornamentals. They have been grown extensively in the southern U.S. as decorative additions to porches, gazebos, walls, and gardens. Most infestations in natural areas are a result of escapes from landscape plantings.

Biology and Spread: Exotic wisterias are long-lived, some vines surviving 50 years or more. Vegetative reproduction is their primary means of expansion. Numerous stolons, or above-ground stems, develop roots and shoots at short intervals. Abundant seeds may also be produced if conditions are favorable, but flower buds are susceptible to winter kill. In riparian habitats, seeds may be carried downstream in water for great distances.

Current Management Approaches: The only practical methods currently available for control of exotic wisterias are mechanical and chemical. Cut climbing or trailing vines as close to the root collar as possible.

This technique, while labor intensive, is feasible for small populations, as a pretreatment for large impenetrable infestations, or for areas where herbicide use is not desirable. Wisteria will continue to resprout after cutting until its root stores are exhausted. For this reason, cutting should begin early in the growing season and, if possible, sprouts cut every few weeks until autumn. Cutting will stop the growth of existing vines and prevent seed production. However, cut vines left coiled around trunks may eventually girdle trees and shrubs as they continue to grow and increase in girth. For this reason, the vines should be removed entirely or at least cut periodically along their length.

Grubbing, removal of entire plants from the roots up, is appropriate for small initial populations or environmentally sensitive areas where herbicides cannot be used. Using a Pulaski, weed wrench or similar digging tool, remove the entire plant, including all roots and runners. Juvenile plants can be hand pulled, depending on soil conditions and root development. Any portions of the root system not removed may resprout. All plant parts (including mature fruit) should be bagged and disposed of in a trash dumpster to prevent reestablishment.

Cut stump treatment, using a systemic herbicide, is effective in areas where vines are established within or around desirable native plants or where they have grown into the canopy. This treatment is effective as long as the ground is not frozen. Cut the stem as close to ground level as possible. Immediately apply a 25% solution of glyphosate (e.g., Roundup) or triclopyr (e.g.,

Garlon) and water to the cross section of the stem. Re-treatment with a foliar application of glyphosate may be necessary for any sprouts.

Use foliar spray herbicide treatments to control large infestations of exotic wisterias. It may be necessary to precede foliar applications with stump treatments to reduce the risk of damaging nontarget species. Apply a 2% concentration of glyphosate (e.g. Round Up) or triclopyr (e.g. Garlon) and water, plus a 0.5% non-ionic surfactant to thoroughly wet all foliage. Chlorpyralid (e.g. Transline) is effective at a concentration of 0.5% and is selective to plants in the aster, buckwheat, and pea families. Caution should be taken with chlorpyralid as groundwater pollution through leaching can be a problem with certain soil types. Do not apply spray so heavily that herbicide drips off the leaves. Glyphosate is a non-selective systemic herbicide that may kill non-target plants that are only partially contacted by spray. Triclopyr is selective to broadleaved species and is a better choice if native grasses are present. Ambient air temperature should be above 65°F for all foliar treatments.

BUFFALO GOURD, STINKING GOURD, CALABACILLA LOCA, MISSOURI GOURD, FOETID WILD PUMPKIN CUCURBITA FOETIDISSIMA KUNTH IN H.B.K.

Buffalo Gourd is a perennial and can be found growing in waste places, fields and along roadsides. It is a low growing vine that can reach a length of 20 feet. The large, yellow flowers open in the mornings and are 3 to 4 inches long and 1 to 2 inches wide. The gray-green leaves are a narrow triangle shape that tend to fold upward parallel to the midvein. The fruit is a gourd that is dark green with lighter green stripes when immature. It is about 2 to 3 inches in diameter at maturity and turns a yellow color. Buffalo Gourd blooms from May to September. It is a perennial, reproducing by seed and roots. Taproot is yellow inside, carrotlike, very thick and fleshy, often more than 6 inches (15 cm) across, sometimes over 5 feet (1.5 m) long. Stems are trailing and rooting at joints, 15 to 30 feet (4.5 to 9 m) long, stout, angled, ridged, rough-hairy, and many-branched. Leaves are thick, simple, sharply toothed, 4 inches to 1 foot (10 to 30 cm) long, ovate, long-pointed, rough above, gray-hairy beneath, with strong, rough-hairy petioles. Flowers are bright yellow, deeply 5-lobed, 3 to 4 inches (7.5 to 10 cm) long and nearly as broad, in the leaf axils.

Male and female flowers are separate but on the same plant. Fruit is globular, 3 inches (7.5 cm) long, with yellow and green markings and a hard, smooth rind. Seeds are numerous, oval, flattened, similar to pumpkin seed but about half as large, and a light straw-color. This plant is found on dry, sandy soil in fields and waste places, and along fence rows. The entire plant has an offensive odor, especially when bruised.

TREES-ALIEN PLANT INVADERS TREE-OF-HEAVEN AILANTHUS ALTISSIMA (MILL.) SWINGLE

Native Range: Central China

Description: Tree-of-heaven, also known as ailanthus, Chinese sumac, and stinking sumac, is a rapidly growing, deciduous tree in the mostly tropical quassia family (Simaroubaceae). Mature trees can reach 80 feet or more in height. Ailanthus has smooth stems with pale gray bark, and twigs which are light chestnut brown, especially in the dormant season. Its large compound leaves, 1-4 feet in length, are composed of 11-25 smaller leaflets and alternate along the stems. Each leaflet has one to several glandular teeth near the base. In late spring, clusters of small, yellow-green flowers appear near the tips of branches. Seeds are produced on female trees in late summer to early fall, in flat, twisted, papery structures called samaras, which may remain on the trees for long periods of time. The wood of ailanthus is soft, weak, coarse-grained, and creamy white to light brown in color. All parts of the tree, especially the flowers, have a strong, offensive odor, which some have likened to peanuts or cashews.

Note: Correct identification of ailanthus is essential. Several native shrubs, like sumacs, and trees, like ash, black walnut and pecan, can be confused with ailanthus. Staghorn sumac (Rhus typhina), native to the eastern U.S., is distinguished from ailanthus by its fuzzy, reddish-brown branches and leaf stems, erect, red, fuzzy fruits, and leaflets with toothed margins.

Ecological Threat: Tree-of-heaven is a prolific seed producer, grows rapidly, and can overrun native vegetation. Once established, it can quickly take over a site and form an impenetrable thicket. Ailanthus trees also produce toxins that prevent the establishment of other plant species. The root system is aggressive enough to cause damage to sewers and foundations.

Distribution in the United States: Tree-of-heaven is widely distributed across the United States, occurring in forty-two states, from Maine to Florida and west to California.

Habitat in the United States: Tree-of-heaven is a common tree in disturbed urban areas, where it sprouts up just about anywhere, including alleys, sidewalks, parking lots, and streets. The book "A Tree Grows in Brooklyn", by Betty Smith is based on the ailanthus tree. Away from cities, it is commonly seen in fields and along roadsides, fencerows, woodland edges and forest openings.

Nationally, ailanthus has become an agricultural pest and may occur as seedlings that pop up by the hundreds in recently planted fields, or as persistent thickets in rocky, untillible areas.

Background: Tree-of-heaven was first introduced to America by a gardener in Philadelphia, PA, in 1784, and by 1840 was commonly available from nurseries. The species was also brought into California mainly by the Chinese who came to California during the gold rush in the mid-1800s. Today it is frequently found in abandoned mining sites there. The history of ailanthus in China is as old as the written language of the country.

Biology and Spread: Tree-of-heaven reproduces both sexually (seeds) and asexually (vegetative sprouts). Flowering occurs late in the spring (June in the middle Atlantic region of eastern U.S.). The species is dioecious, with male and female flowering on separate trees. Fruits are papery, somewhat twisted, winged structures called samaras that are tan to pinkcolored. Samaras occur in large clusters from September to October of the same year, and may persist on the tree through the following winter. One study reports that an individual tree can produce as many as 325,000 seeds per year. Established trees also produce numerous suckers from the roots and resprout vigorously from cut stumps and root fragments.

Current Management Approaches: Elimination of Ailanthus requires diligence, due to its abundant seed production, high seed germination rate, and vegetative reproduction. Follow-up monitoring and treatment when needed should be an integral part of any serious ailanthus management program. Regardless of method selected, treated areas should be rechecked one or more times a year and any new suckers or seedlings treated (cut, sprayed or pulled) as soon as possible, especially before they are able to rebuild root reserves. Establishing a thick cover of trees (preferably native and non-invasive) or grass sod will help shade out and discourage establishment of ailanthus seedlings. Targeting large female trees for control will help reduce spread of ailanthus by seed.

Young seedlings may be pulled or dug up, preferably when soil is moist. Care must be taken to remove the entire plant, including all roots and fragments, as these will almost certainly re-grow. Root suckers appear similar to seedlings, but would be connected to a pre-existing lateral root, and would be nearly impossible to remove effectively.

Cutting alone is usually counter-productive because ailanthus responds by producing large numbers of stump sprouts and root suckers. However, for small infestations, repeated cutting of sprouts over time can exhaust the plants reserves and may be successful if continued for many years or where heavy shade exists. If possible, the initial cutting should be in early summer in order to impact the tree when its root reserves are lowest. Cutting large seed producing female trees would at least temporarily reduce spread by this method.

The most effective method of ailanthus control seems to be through the use of herbicides, which may be applied as a foliar (to the leaves), basal bark, cut stump, or hack and squirt treatment. Keep in mind that it is relatively easy to kill the above ground portion of ailanthus trees--you need to kill or seriously damage the root system to prevent or limit stump sprouting and root suckering. Always be extremely careful with herbicide applications in the vicinity of valuable ornamental shrubs and trees. Foliar sprays applied when trees are in full leaf are very effective, and should be the method of choice where ailanthus size and distribution allows effective spray coverage of all foliage without unacceptable contact with nearby desirable vegetation or the applicator. Where ailanthus is in association with other exotic weed species, as is often the case, foliar spray allows treatment of the entire area at one time. Limitations of the method are the seasonal time frame, the need to transport a larger, more diluted volume of spray material, and the fact that rapid growing ailanthus are often out of effective reach.

The non-selective herbicide glyphosate (e.g., Roundup, Rodeo, Accord), will kill or injure almost any plant, herbaceous or woody, contacted by the spray. Triclopyr (e.g., Garlon 3A, Garlon 4) is selective for broadleaf and woody plants and will not kill grasses contacted by the spray. Both glyphosate and triclopyr are systemic herbicides, meaning that they are absorbed by plants and are carried to the root systems. These herbicides have low soil activity, so do not pose a threat to groundwater if applied properly and at recommended label rates. Both glyphosate and triclopyr should be mixed with water and a small amount (0.5%, or as per label) of a non-ionic surfactant (except for Roundup, which contains a surfactant) to help the spray spread over and penetrate the leaves. The mixture should be applied to leaves and green stems, including sprouts and suckers, until thoroughly wet but not to the point of runoff. With backpack sprayers, concentrations of 2% of a typical glyphosate product such as Roundup or Accord applied June 15-September 15, or 1.5% of a 4 lb./gallon triclopyr product such as Garlon 4, or 2% of a 3 lb./gallon triclopyr product such as Garlon 3A applied June 1-September 1 have worked well in the Mid-Atlantic area, with slightly greater effectiveness for the triclopyr products. For higher volume applications such as would be applied by a truck mounted sprayer, the concentration for these products could be reduced by 0.5% to 1-1.5%. Other herbicides which have shown to be effective for foliar application of ailanthus are dicamba (e.g., Banvel, Vanquish), imazapyr (e.g., Arsenal, Chopper), and metsulfuron methyl (e.g., Escort).

Basal bark application is one of the easiest methods and does not require any cutting. It works best during late winter/ early spring and in summer. The base of the tree stem must be free of snow, ice, or water on the bark from recent rainfall, though precipitation following application is inconsequential. Late winter/early spring (February 15-April 15, Mid-Atlantic) is generally the most productive time, since vegetation near the base of the trees is usually absent or leafless. Late spring and early summer applications (April 15-June 1, Mid-Atlantic), when plant fluids are moving upwards to support new growth, are questionable. Application during the summer (June 1-September 15, Mid-Atlantic) works very well as long as vegetation is not a hindrance, and allows lower concentrations of herbicide to be used. Fall to mid-winter applications (October-January) have given poor results. Mix up a solution of 20% (as low as 10% in summer depending on objectives) concentration of oil-soluble triclopyr product (e.g., Garlon 4) in 80% oil (fuel oil, diesel, kerosene, mineral oil, or special vegetable oils). With these diluents some applicators add a pine oil based additive (e.g., Cide-Kick II) at the rate of 10%, which helps penetrate the bark and eliminate any unpleasant odor. Some companies market diluents based on mineral or vegetable oils specifically designed for basal bark application, which should be considered for use in sensitive areas. Another option is to use a pre-mixed, ready-to-use triclopyr product designed for basal bark (and cut stump) application (e.g., Pathfinder II). Using a handheld or backpack type sprayer, apply the mixture in a 12 inch wide band around the entire circumference of the tree base with no "skips". The basal bark method is generally used for trees that are less than 6 inches in diameter, though slightly larger stems may also be treated effectively by thoroughly treating bark up to 24 inches in height.

Follow-up foliar herbicide application (see above) to basal sprouts and root suckers may be necessary. Another herbicide which has been shown to be effective for basal bark control of ailanthus is imazapyr (e.g., Chopper, Stalker). This is sometimes used in a combination with triclopyr at a concentration of 15% Garlon 4 and 5% Stalker in 80% oil diluent. Root suckering will be an increasing problem in the fall, winter and spring. The hack-and-squirt (or injection) method is very effective and minimizes sprouting and suckering when applied during the summer. This method requires first making downward-angled cuts into the sapwood around the tree trunk at a comfortable height, using a hand ax. With spray bottle or wand in the other hand, squirt a straight (100%) concentration of a water-soluble triclopyr product (e.g., Garlon 3A) into the cuts within a minute or two, applying 1-2 milliliters into each cut (typically 1-2 squirts of a trigger squirt bottle) so that the bottom of the cut is covered, but liquid doesn't run out of it. Generally, you would make about 1 hack cut for each inch of diameter plus one (i.e., for a 10 inch diameter tree, make about 11 cuts). Space the cuts so that about 1-2 inches of uncut living tissue remains between them. A continuous line of cuts around the trunk would likely cause the tree to go into emergency response mode and react by producing basal sprouts and root suckers. For this reason, girdling or frilling (girdling followed by herbicide) is not highly recommended unless long term follow-up treatment is possible. While spaced injection works well for ailanthus, it is not as effective on some other species. This method can be used with trees of any size, though it is most productive with stems over 2 inches in diameter. This method is relatively easy for one person to do, with hatchet in one hand and spray bottle in the other, but should be done with a buddy nearby in case of an accident. Monitor the treatment area and be prepared to follow-up with a foliar application the next year to control any basal sprouts or root suckers that might emerge. Glyphosate products have sometimes been recommended for control of ailanthus using this method, but several field trials have shown consistently poor longterm control of basal sprouts and root suckers at any time of year. Other herbicides which have shown to be effective for hack-and-squirt control of ailanthus during the growing season are dicamba (e.g., Banvel, Vanquish), imazapyr (e.g., Arsenal A.C., Chopper), and 2,4-D + picloram (e.g., Pathway). Dicamba is particularly effective in October.

The cut stump method is useful in areas where the trees need to be removed from the site and will be cut as part of the process. While situations exist that dictate this method over the others given above, felling trees is usually less effective in killing the root system, slower, more labor intensive, and more hazardous to personnel than other methods. This method is likely to be most successful during the growing season, with diminishing success through the early fall. Dormant season applications may prevent resprouting from the stump itself, but will do little to inhibit root suckering. However, at any time of year, if the tree must be cut it is better to treat the stump than not. Application of herbicide to the cut stumps must be conducted immediately after cutting, within 5-15 minutes of the cut with water soluble formulations, longer with oil mixtures, to ensure uptake of the chemical before the plant seals the cut area off. The mixture may be painted on with a paint brush or sprayed on using a spray bottle or backpack sprayer. A mixture of 20% Garlon 4 plus 80% oil diluent, as for basal bark spraying (above), may be used. In this case the whole stump surface and sides to the ground line would be sprayed. Another option is to use Garlon 3A at 100%, treating only the outer 1/3 of the stump surface. Be prepared to follow-up with a foliar application the next year to control any stump sprouts or root suckers which emerge. Other herbicides which have shown to be effective in stump treatment of ailanthus are the same as those listed above for hack and squirt or injection. A potential biological control for ailanthus may lie in several fungal pathogens, (Verticillium dahliae and Fusarium oxysporum) that have been isolated from dead and dying ailanthus trees in New York and in southern and western Virginia.

SILK TREE ALBIZIA JULIBRISSIN DURZ.

Native Range: Iran to Japan (according to Hortus 3rd. ed.).

Description: Silk tree, also known as mimosa, or silky acacia, is a small to medium-sized tree in the pea family (Fabaceae) that can grow up to 20-40 feet tall. The bark is light brown, nearly smooth, and generally thin with lens shaped areas along the stem. The attractive fern-like leaves of mimosa are finely divided, 5-8 inches long by about 3-4 inches wide, and alternate along the stems. Silk tree has showy and fragrant pink flowers, about 1½ inches long, that resemble pom-poms and are arranged in panicles at the ends of branches. Fruits are flat, straw-colored pods about 6 inches long containing light brown ovalshaped seeds about ½ inch in length. Pods ripen in August to September and begin to disintegrate soon after, but remain on the trees into winter.

Ecological Threat: Because silk tree can grow in a variety of soils, produce large seed crops, and resprout when damaged, it is a strong competitor to native trees and shrubs in open areas or forest edges. Dense stands of mimosa severely reduce the sunlight and nutrients available for other plants.

Distribution in the United States: Silk tree is naturalized from New Jersey to Louisiana and in California.

Habitat in the United States: Silk tree takes advantage of disturbed areas, often spreading by seed from nearby ornamentals or from contaminated fill dirt. It prefers full sun and is often seen along roadsides and open vacant lots in urban/suburban areas. Silk tree can tolerate partial shade but is seldom found in forests with full canopy cover, or at higher elevations (above 900 m or 3,000 ft), where cold hardiness is a limiting factor. It can, however, become a serious problem along riparian areas, where it becomes established along scoured shores and where its seeds are easily transported in water. Like many successful exotics, it is capable of growing in a wide range of soil conditions.

Background: Silk tree was introduced to the U.S. in 1745. Silk tree continues to be a popular ornamental because of its fragrant and showy flowers.

Biology and Spread: Silk tree reproduces both vegetatively and by seed. Silk tree seeds have impermeable seed coats that allow them to remain dormant for many years. One study showed that 90% of the seeds were viable after five years and, for another species of mimosa, a third of its seeds germinated after 50 years in open storage. Seeds are mostly dispersed below or around the parent plant, but can be dispersed further by water. Silk trees grow rapidly under good conditions but are short-lived and have weak, brittle wood. If cut or top-killed, trees resprout quickly and sprouts can grow over three feet in a season.

Current Management Approaches: Silk tree can be controlled using a variety of mechanical and chemical controls. Trees can be cut at ground level with power or manual saws. Cutting is most effective when trees have begun to flower to prevent seed production. Because mimosa spreads by suckering, resprouts are common after treatment. Cutting is an initial control measure and will require either an herbicidal control or repeated cutting for resprouts.

Girdling is effective on large trees where the use of herbicides is impractical. Using a hatchet, make a cut through the bark encircling the base of the tree, approximately six inches above the ground. Be sure that the cut goes well below the bark. This method will kill the top of the tree but resprouts are common and may require a follow-up treatment with a foliar herbicide.

Plants should be pulled as soon as they are large enough to grasp, but before they are old enough to flower. Seedlings are best pulled after a rain when the soil is loose. The entire root must be removed since broken fragments may resprout. Silk tree seedlings and small trees can be controlled by applying a 2% solution of glyphosate (e.g., Roundup) or triclopyr (e.g., Garlon) and water plus a 0.5% non-ionic surfactant to thoroughly wet all leaves. Systemic herbicides such as glyphosate and triclopyr can kill entire plants because the chemicals travel through a plant from the leaves and stems to the actively growing roots, where they prevent further cell growth. Use a low pressure and a coarse spray pattern to reduce damage from spray drift on non-target species. Use caution when applying these products, as glyphosate is a nonselective herbicide that may kill non-target plants that are only partially contacted. Triclopyr is a selective herbicide for many broadleaf plant species. Triclopyr is a selective herbicide for many broad-leaved plant species and should be considered for sites where native or other desirable grasses are meant to be conserved.

The cut-stump and basal bark herbicidal methods should be considered when treating individual trees or where the presence of desirable species preclude foliar application. Stump treatments can be used as long as the ground is not frozen. Horizontally cut stems at or near ground level. And immediately apply a 25% solution of glyphosate or triclopyr and water to the cut stump, making sure to cover the outer 20% of the stump. Basal bark applications are effective throughout the year as long as the ground is not frozen. Apply a mixture of 25% triclopyr and 75% horticultural oil to the base of the tree trunk to a height of 12-15 inches from the ground. Thorough wetting is necessary for good control; spray until run-off is noticeable at the ground line.

AUSTRALIAN PINE CASUARINA EQUISETIFOLIA L.

Native Range: Malaysia, southern Asia, Oceania and Australia.

Description: A member of the beefwood family (Casuarinaceae), Australian pine is a deciduous tree with a soft, wispy, pine-like appearance that can grow to 100 feet or more in height. Also known as ironwood, beefwood, she oak and horsetail tree, it bears a superficial resemblance to the conifer genus Pinus because of its small, round, cone-like fruits and its branchlets of scale-like leaves that look like pine needles. Its flowers are tiny, brown and windpollinated. The fruit is a nutlet about ½ inch in diameter that contains winged seeds.

Ecological Threat: Australian pine is fastgrowing (5-10 feet per year), produces dense shade and a thick blanket of leaves and hard, pointed fruits that completely covers the ground beneath it. Dense thickets of Australian pine displace native dune and beach vegetation, including mangroves and many other resident, beach-adapted species. Because its roots are capable of producing nitrogen through microbial associations, Australian pine can colonize nutrient-poor soils. Once established, it radically alters the light, temperature, and soil chemistry regimes of beach habitats, as it outcompetes and displaces native plant species and destroys habitat for native insects and other wildlife.

Chemicals in the leaves of Australian pine may inhibit the growth of other plants underneath it. The ground below Australian pine trees becomes ecologically sterile and lacking in food value for native wildlife. Unlike native shrubbery, the thick, shallow roots of Australian pine make it much more susceptible to blow-over during high wind events, leading to increased beach and dune erosion and interference with the nesting activities of sea turtles.

Distribution in the United States: Australian pine is established in the Hawaiian islands, coastal Florida, Puerto Rico, the Bahamas, and many Caribbean islands.

Habitat in the United States: Australian pine occurs in open, coastal strand habitat, characterized by sand and shell beaches, rocky coasts, sand dunes, and sand bars in subtropical climates.

Biology and Spread: Australian pine is capable of flowering year-round and can produce tremendous numbers of small, winged seeds which are dispersed by wind.

Background: Australian pine was introduced to Florida in the late 1800's and planted widely for the purposes of ditch and canal stabilization, shade and lumber.

Current Management Approaches: No biological controls are currently available for management of Australian pine. For new or small infestations, manual removal of Australian pine seedlings and saplings is recommended. For heavier infestations, application of a systemic type herbicide to bark, cut stumps, or foliage is likely to be the most effective management tool. Prescribed fire has also been used for large infestations in fire-tolerant communities. Raking and removal of leaf litter, cones and seeds should be done whenever possible. Impacts to native plants should be minimized during any control activities. Whenever possible, efforts should be taken to prevent the introduction or encroachment of Australian pine. For example, recently disturbed beach habitat may be planted with native vegetation to prevent Australian pine from invading.

CARROTWOOD CUPANIOPSIS ANACARDIOIDES (A. RICH.) RADLKF.

Native Range: Australia, Iran Jaya (Indonesia) and Papua New Guinea

Description: Carrotwood is a fastgrowing evergreen tree in the Soapberry family (Sapindaceae) that grows to a height of about 35 feet. The leaves are large and compound, made up of four to ten oblong leaflets, each 4 to 8 inches long, and attached by a swollen stalk. Leaflet edges tend to be wavy with rounded tips that are often indented. Leaves alternate along the stems. In Florida, flowering occurs in the winter, from January to March. Clusters of small, greenish-white flowers are borne on stalks that emerge from leaf axils. Flowers are unisexual, with each flower cluster containing both male and female flowers. The brightly colored fruit is a yellow, three-lobed capsule which, when ripe (May to June) splits open to expose three shiny black seeds encased in red or orange fleshy tissue.

Ecological Threat: While carrotwood invades a variety of natural communities, including dunes, coastal strand, sand pine scrub, slash pine flatwoods, cypress swamps, freshwater marshes and river banks, it poses a special threat to coastal ecosystems like mangrove swamps and tropical hammocks. Coastal plant communities provide crucial erosion control, water quality benefits, and food and shelter for wildlife. Once introduced, carrotwood forms dense monocultures, crowding out and out-competing native plants for available light and nutrients. Because mangroves provide critical habitat for wading and diving birds, some of which are designated Species of Special Concern, and serve as nursery grounds for crabs, other crustaceans, invertebrates and commercial and recreational fish, the impacts of carrotwood establishment are serious and farreaching. Coastal hammocks and mangroves are continually losing ground to development and are also impacted by natural forces such as tropical storms and hurricanes. Alteration of species compcsition and competition by invasive exotic species increases stress to the remaining hammocks. Because carrotwood is a popular, fast-growing landscape tree that is widely planted and very adaptable, the impacts to mangroves and other habitats are expected to increase. Carrotwood has also been found growing among other aggressive, invasive exotic trees.

Distribution in the United States: As of 1996, carrotwood has been documented to occur in natural areas in fourteen Florida counties, from Brevard and Hillsborough counties southward. The current distribution of carrotwood parallels that of mangrove tree species. While naturalized carrotwood infestations are limited primarily to coastal areas, inland populations are beginning to surface. Carrotwood has also been used ornamentally in California, but there are no reports of naturalized populations there, perhaps due to their drier climate. Cold tolerance may limit its potential distribution. According to one reference, carrotwood is able to withstand temperatures to about 22 F (-6 C). Test specimens in northern Florida, however, have withstood winters at least that cold.

Habitat in the United States: Tolerant of salt, poor soils, poor drainage, sunlight and shade, carrotwood can adapt to dry areas, and appears in disturbed and undisturbed sites. As a result, carrotwood inhabits a variety of habitats including coastal hammocks, dunes, coastal strand, sand pine scrub, slash pine flatwoods, mangrove swamps, cypress swamps, freshwater marshes and river banks.

Background: University of Florida Herbarium specimens document carrotwood cultivation as early as 1955 in eastern Florida. A separate introduction in Sarasota, Florida in 1968 resulted in large scale propagation and use as an ornamental tree.

Carrotwood became a popular landscape tree throughout southern Florida in the late 1970s and early 1980s. By 1990, wild carrotwood seedlings began to be seen in the wild in various habitats.

Biology and Spread: Carrotwood is a prolific seed producer, and the brightly colored fruits are very attractive to birds which disperse it widely. Bird dispersal explains isolated island populations and seedlings under trees and telephone poles. Seedlings have also been found along estuary rack lines. Clumps of seedlings suggest dispersal by small mammals. In its native range, carrotwood is pollinated by bees, which are the likely pollinators in Florida.

Current Management Approaches: No biological control is available at this time. Chemical control is the most common and effective method of control. Triclopyr (e.g., Garlon) has proved effective as a basal bark treatment and cut stump treatment. Glyphosate (e.g., Rodeo) is marginally successful, and usually requires retreatment. Care must be taken in mangrove and wetland areas to avoid impacts to sensitive flora and fauna by use of chemicals or heavy equipment. As a preventive measure, a few counties and municipalities have ordinances restricting use of carrotwood.

PRINCESS TREE PAULOWNIA TOMENTOSA (THUNB.) SIEB. & ZUCC. EX STEUD.

Native Range: China

Description: Princess tree, also known as royal paulownia or empress tree, is a small to medium sized tree in the figwort family (Scrophulariaceae) that may reach 30-60 feet in height. The bark is rough, graybrown, and interlaced with shiny, smooth areas. Stems are olive-brown to dark brown, hairy and markedly flattened at the nodes (where stems and branches meet). Leaves are large, broadly oval to heart-shaped, or sometimes shallowly three-lobed, and noticeably hairy on the lower leaf surfaces.

They are arranged in pairs along the stem. Conspicuous upright clusters of showy, pale violet, fragrant flowers open in the spring. The fruit is a dry brown capsule with four compartments that may contain several thousand tiny winged seeds. Capsules mature in autumn when they open to release the seeds and then remain attached all winter, providing a handy identification aid.

Ecological Threat: Princess tree is an aggressive ornamental tree that grows rapidly in disturbed natural areas, including forests, streambanks, and steep rocky slopes.

Distribution in the United States: Princess tree is found in 25 states in the eastern U.S., from Maine to Texas.

Habitat in the United States: Princess tree can be found along roadsides, streambanks, and forest edges. It tolerates infertile and acid soils and drought conditions. It easily adapts to disturbed habitats, including previously burned areas, forests defoliated by pests (such as the gypsy moth) and landslides and can colonize rocky cliffs and scoured riparian zones where it may compete with rare plants in these marginal habitats. Its ability to sprout prolifically from adventitious buds on stems and roots allows it to survive fire, cutting, and even bulldozing in construction areas.

Background: Princess tree was introduced into the U.S. as an ornamental and landscape tree around 1840. It was first imported to Europe in the 1830's by the Dutch East India Company and brought to North America a few years later. This tree has since become naturalized in the eastern U.S. and is also grown on the west coast. Princess tree is native to western and central China where historical records describe its medicinal, ornamental, and timber uses as early as the third century B.C. It was cultivated centuries ago in Japan where it is valued in many traditions. Recently it has also been grown in plantations and harvested for export to Japan where its wood is highly valued.

Biology and Spread: Princess tree can reproduce from seed or from root sprouts; the latter can grow more than 15 feet in a single season. The root branches are shallow and horizontal without a strong taproot. Seed-forming pollen is fully developed before the onset of winter and the insectpollinated flowers open in spring. A single tree is capable of producing an estimated twenty million seeds that are easily transported long distances by wind and water and may germinate shortly after reaching suitable soil. Seedlings grow quickly and flower in 8-10 years. Mature trees are often structurally unsound and rarely live more than 70 years.

Current Management Approaches: Princess trees can be controlled using a variety of mechanical and chemical controls. Hand pulling may be effective for young seedlings. Plants should be pulled as soon as they are large enough to grasp. Seedlings are best pulled after a rain when the soil is loose. The entire root must be removed, since broken fragments may resprout. Trees can be cut at ground level with power or manual saws. Cutting is most effective when trees have begun to flower to prevent seed production. Because Princess tree spreads by suckering, resprouts are common after cutting.

Cutting should be considered an initial control measure that will require either repeated cutting of resprouts or an herbicidal treatment. Princess tree seedlings and small trees can be controlled by applying a 2% solution of glyphosate (e.g., Roundup) or triclopyr (e.g., Garlon) and water plus a 0.5% non-ionic surfactant to thoroughly wet all leaves. Use a low pressure and coarse spray pattern to reduce damage from spray drift on non-target species. Glyphosate is a non-selective systemic herbicide that may kill non-target plants that are only partially sprayed. Triclopyr is a selective herbicide for broadleaf species. In areas where desirable grasses are growing, triclopyr can be used with minimal non-target damage.

Girdling is effective on large trees where the use of herbicides is impractical. Using a hatchet, make a cut through the bark encircling the base of the tree, approximately six inches above the ground. Be sure that the cut goes well below the bark. This method will kill the top of the tree but resprouts are common and may require a follow-up treatment with a foliar herbicide. The cut stump method (that is, applying herbicide to freshly cut stumps), should be considered for individual trees or when desirable plants are nearby that might be impacted by foliar applications. Stump treatments can be used as long as the ground is not frozen. Begin treatments by horizontally cutting stems at or near ground level. Immediately apply a 50% solution of glyphosate or triclopyr and water to the cut stump making sure to cover the outer 20% of the stump. Basal bark applications are effective throughout the year as long as the ground is not frozen. Apply a mixture of 25% triclopyr and 75% horticultural oil to the base of the tree trunk to a height of 12-15 inches from the ground. Thorough wetting is necessary for good control; spray until run-off is noticeable at the ground line. For more information on the management of Princess Tree, please contact:

WHITE POPLAR POPULUS ALBA L.

Native Range: Central and southern Europe to western Siberia and central Asia.

Description: White poplar, also known as silver-leaved or silverleaf poplar, is a tall member of the willow family (Salicaceae) that, at maturity, may reach 70 feet or more in height and 2 feet in diameter. The smooth, greenish-white bark becomes dark and rough on older trees. Young green or brown twigs are coated with dense woolly hair, especially near the tip. A cross-section of the stem reveals a five-pointed, star-shaped pith.

The 2 to 5-inch long leaves are oval to maple-leaf in shape with 3-5 broad teeth or lobes, and are dark green above and covered with dense white hair below. Male and female flowers are borne in catkins on separate trees and appear sometime in March and April. The small seeds are adorned with cottony fluff that is easily blown by the wind in late spring, and is a bane to many landscape maintenance workers.

Ecological Threat: White poplar outcompetes many native tree and shrub species in mostly sunny areas, such as forest edges and fields, and interferes with the normal progress of natural community succession. It is an especially strong competitor because it can grow in a variety of soils, produce large seed crops, and resprouts easily in response to damage. Dense stands of white poplar prevent other plants from coëxisting by reducing the amount of sunlight, nutrients, water and space available.

Distribution in the United States: White poplar is found in forty-three states throughout the contiguous U.S.

Habitat in the United States: White poplar seems to grow best in full sun habitats such as fields, forest edges and wetland fringes.

Background: White poplar was first introduced to North America in 1748 and has a long history in cultivation. It is chiefly planted as an ornamental for its attractive leaves of contrasting color (i.e., green above, white below). It has escaped and spread widely from many original planting sites. Because it is susceptible to a wide variety of pest insects and diseases, and is easily damaged by storms and wind, the ornamental value of white poplar is low.

Biology and Spread: Local spread of white poplar is primarily by vegetative means, through root suckers. Root suckers arise from adventitious buds on the extensive lateral root system. Large numbers of suckers from a single tree can quickly develop into a dense colony. Suckering can occur naturally or as a result of damage or other disturbance to the parent plant. Mature white poplar trees produce thousands of wind-dispersed seeds that may be carried long distances. However, seed germination of white poplar appears to be very low in the U.S.

Current Management Approaches: White poplar can be controlled using a variety of physical and chemical controls. Removal of seedlings and young plants by hand will help prevent further spread or establishment. Plants should be pulled as soon as they are large enough to grasp. The entire root system, or as much of it as possible, should be removed to prevent resprout from fragments. Hand removal of plants is best achieved after a rain, when the soil is loose. Trees of any size may be felled by cutting at ground level with power or manual saws. Because resprouts are common after cutting, this process may need to be repeated many times until the reserves of the tree are exhausted. Girdling, which kills the tree by severing tissues that conduct water and sugars, also may be effective for large trees, especially if accompanied by application of a systemic herbicide to the cut area.

A hatchet or saw is used to make a cut through the bark encircling the base of the tree, approximately six inches above the ground and deep into the bark. Girdling will kill the parent tree but may require follow-up cutting or treatment of sprouts with an herbicide. Chemical control of white poplar seedlings and small trees has been achieved by applying a 2% solution of glyphosate (e.g., Roundup) or triclopyr (e.g., Garlon 3) and water plus a 0.5% non-ionic surfactant to the foliage until the leaves are thoroughly wet. Use of low pressure and a coarse spray with large droplet size will reduce spray drift and damage to non-target plants.

Caution: Because glyphosate is a non-selective systemic herbicide, it may kill other grasses, broadleaf herbaceous and woody plants that it contacts. Triclopyr kills broadleaf (dicotyledonous) plants but causes little or no damage to grasses and is useful for areas where desirable grasses are to be maintained. The cut stump herbicidal method should be considered when treating individual trees or where the presence of desirable species precludes the use of foliar herbicides. Stump treatments can be made at any time of year as long as the ground is not frozen. After cutting the tree near ground level, a 25% solution of glyphosate or triclopyr and water is applied to the stump by spray bottle or brush, making sure to cover the outer 20% of the stump. Basal bark herbicidal treatment is also effective throughout the year, as long as the ground is not frozen, and does not require cutting of the tree. A mixture of 25% triclopyr in an ester formulation (e.g., Garlon 4) and 75% horticultural oil is applied to the bark in a wide band around the base of the tree to a height of 12-15 inches from the ground. Thorough wetting is necessary for good control; spray until run-off is just noticeable at the ground line, but not running off-site.

COMMON BUCKTHORN RHAMNUS CATHARTICA L.

Native Range: Eurasia

Description: Common buckthorn is a shrub or small tree in the buckthorn family (Rhamnaceae) that can grow to 22 feet in height and have a trunk up to 10 inches wide. The crown shape of mature plants is spreading and irregular. The bark is gray to brown, rough textured when mature and may be confused with that of plum trees in the genus Prunus. When cut, the inner bark is yellow and the heartwood, pink to orange. Twigs are often tipped with a spine. In spring, dense clusters of 2 to 6, yellow-green, 4-petaled flowers emerge from stems near the bases of leaf stalks. Male and female flowers are borne on separate plants. Small black fruits about 1/4 inch in cross-section and containing 3-4 seeds, form in the fall. Leaves are broadly oval, rounded or pointed at the tip, with 3-4 pairs of up curved veins, and have jagged, toothed margins. The upper and lower leaf surfaces are without hairs. Leaves appear dark, glossy green on the upper surface and stay green late into fall, after most other deciduous leaves have fallen.

A similar problem exotic species is Rhamnus frangula, glossy buckthorn. Glossy buckthorn does not have a spine at twig tips, leaves are not toothed, and the undersides of the leaves are hairy.

Several native American buckthorns occur in the eastern U.S. that could be confused with the exotic species. If in doubt, consult with a knowledgeable botanist to get an accurate identification. Carolina buckthorn (Rhamnus caroliniana), is a lovely native shrub that has finely toothed leaves somewhat resembling those of black cherry, and are smooth on the underside; it produces attractive fruits from August to October. Alder buckthorn (Rhamnus alnifolia), is a low-growing shrub that may grow to a maximum of 3 feet in height, and has leaves with 6-7 pairs of veins.

Ecological Threat: Exotic buckthorns tend to form dense, even-aged thickets, crowding and shading out native shrubs and herbs, often completely obliterating them. Dense buckthorn seedlings prevent native tree and shrub regeneration. In fire-adapted ecosystems such as savannas and prairies, the lack of vegetation under buckthorn prohibits fires. Buckthorn control is also of interest to small grain producers; the shrub is an alternate host of the crown rust of oats, which affects oat yield and quality.

Distribution in the United States: Common buckthorn has become naturalized from Nova Scotia to Saskatchewan, south to Missouri, and east to New England.

Habitat in the United States: Common buckthorn prefers lightly shaded conditions. An invader mainly of open oak woods, deadfall openings in woodlands, and woods edges, it may also be found in prairies and open fields. It is tolerant of many soil types, well drained sand, clay, poorly drained calcareous, neutral or alkaline, wet or dry.

Background: Common buckthorn was introduced to North America as an ornamental shrub, for fence rows, and wildlife habitat. Introduction of buckthorn was based on its hardiness and ability to thrive in a variety of soil and light conditions.

Biology and Spread: Common buckthorn is a dioecious plant, meaning that each plant produces only male or female flowers and fruiting trees are always female. Most of the fruits fall directly beneath the shrubs, creating a dense understory of seedlings characteristic of common buckthorn stands. The plentiful fruit is eaten by birds and mice and is known to produce a severe laxative effect, helping distribute seeds through birds, often far from the parent plant. Buckthorn often establishes beneath trees at the edges of forests and fields.

Current Management Approaches: Mechanical, physical and chemical methods are available for control of common buckthorn and glossy buckthorn (Rhamnus frangula), also an invasive exotic plant. Prescribed fire is one method proposed for controlling buckthorn seedlings in fire-adapted natural areas, from late March to early May, most recently by Boudreau and Wilson. In the upper Midwest conduct burns as soon as leaf litter is dry; resprouts will be less vigorous due to low carbohydrate levels. Burning every year or every other year in established stands may be required for 5-6 years or more. Unfortunately, buckthorn seedlings often grow in low litter areas, unsuitable for frequent prescribed fire. In dense stands, seedlings and saplings may be cut and dropped on site, creating fuel for future fires. Buckthorn seedlings appear vulnerable to fire, perhaps due to their poorly established root structure. Fire will top kill a mature plant, but resprouting does occur. Uprooting of 1/2 inch diameter seedlings by hand or up to 1 1/2 inch diameter using a weed wrench is effective, but care should be taken to avoid excessive disturbance to the soil, which can release buckthorn seeds stored in the soil.

Careful application of herbicides has been found to effectively control buckthorn in Illinois. The McHenry County, Illinois, Conservation District (MCICD) reports excellent results using a triclopyr herbicide at the rate of 1:4 herbicide:water with dye on cut stumps during the growing season, from late May to October. The product label suggests avoiding treatment during the spring sap flow. To extend the work season, the use of a triclopyr herbicide was also applied to cut stumps during winter and was reported to be effective by MCICD and the Minnesota Region V State Parks. Frill application (applying herbicide into the cambial layer of fresh cuts on the tree trunk) using the 1:4 rate of triclopyr herbicide with oil and dye was also effective.

Experiments at the University of Wisconsin Arboretum report good results using a mixture of 1 part triclopyr herbicide to 7 parts oil on cut stumps, or a 1 part triclopyr herbicide to 16 parts oil mixture applied as a basal bark treatment to stems less than 3 inches across. For fall applications, the Minnesota Department of Natural Resources, Region V State Parks Resource Management has used a 1 part glyphosate herbicide to 5 parts water mixture applied immediately to cut stumps using a hand sprayer. Initial checks indicated over 85 percent control at the test site.

SALTCEDAR TAMARIX APHYLLA, T. CHINENSIS, T. GALLICA, T. PARVIFLORA, AND T. RAMOSISSIMA

Native Range: Eurasia and Africa

Description: Most saltcedars, or tamarisks, are deciduous shrubs or small trees growing to 12-15 feet in height and forming dense thickets. Tamarix aphylla is an evergreen tree that can grow to 50 feet tall and tends to flower during the winter. Saltcedars are characterized by slender branches and gray-green foliage. The bark of young branches is smooth and reddish-brown. As the plants age, the bark becomes brownishpurple, ridged and furrowed. Leaves are scalelike, about 1/16 inch long and overlap each other along the stem. They are often encrusted with salt secretions. From March to September, large numbers of pink to white flowers appear in dense masses on 2-inch long spikes at branch tips. Common Name: saltcedar

Scientific Names: (deciduous) Tamarix ramosissima Ledelbour T. pentandra Pallas T. chinensis Loureiro T. parviflora DeCandolle T. gallica L. (evergreen) T. aphylla (L.) Karsten Family: Tamaricaceae

Taxonomy: Tamarix ramosissima is a Class A noxious weed in many states, and so eradication is required by state law. However, it is very difficult to distinguish one species of saltcedar from another and there is taxonomic disagreement of the deciduous species. They are difficult to distinguish and apparently they have the ability to hybridize. At various times, the deciduous species have been grouped together under T. ramosissima, T. chinensis or T. pentandra, depending on the researcher. The most recent school of thought has this group divided into:

T. parviflora-flowers with 4 sepals and 4 petals

T. ramosissima-flowers with 5 sepals and 5 petals

Comparison of Differences between T. parviflora and T. ramosissima

Plant Characteristics Species Species

T. parviflora T. ramosissima

Bark dark brown to deep purple reddish brown

Flowering Stems 1-1.5 inch long 3-4 inches long

Flower Parts 4 sepals and 4 petals 5 sepals and 5 petals

Petal Size 1.9-2.3 mm long 1-1.8 mm long

Flower Color reddish pale pink to whitish

Flowering Time early May to June-Flowers appear before leafing out

May to Oct. until freeze or water loss-Flowers appear after leafing out

Seed Production about one month all growing season

Ecological Threat: Saltcedars are fire-adapted species and have long tap roots that allow them to intercept deep water tables and interfere with natural aquatic systems. Saltcedar disrupts the structure and stability of native plant communities and degrades native wildlife habitat by outcompeting and replacing native plant species, monopolizing limited sources of moisture, and increasing the frequency, intensity and effect of fires and floods. Although it provides some shelter, the foliage and flowers of saltcedar provide little food value for native wildlife species that depend on nutrient-rich native plant resources.

Distribution in the United States: Saltcedar occurs in the intermountain region of the western United States, throughout the Great Basin, and California and Texas.

Habitat in the United States: Saltcedar establishes in disturbed and undisturbed streams, waterways, bottomlands, banks and drainage washes of natural or artificial waterbodies, moist rangelands and pastures, and other areas where seedlings can be exposed to extended periods of saturated soil for establishment. Saltcedar can grow on highly saline soils containing up to 15,000 ppm soluble salt and can tolerate alkali conditions.

Background: Saltcedar was introduced to the western U.S. as an ornamental shrub in the early 1800s.

Biology and Spread: Saltcedar spreads vegetatively, by adventitious roots or submerged stems, and sexually. Each flower can produce thousands of tiny (1/25-inch diameter) seeds that are contained in a small capsule usually adorned with a tuft of hair that aids in wind dispersal. Seeds can also be dispersed by water. Seedlings require extended periods of soil saturation for establishment.

Current Management Approaches: Management of saltcedar requires a long term commitment to maintain at low levels and prevent reinfestation. A variety of methods have been used in the management of saltcedar, including mechanical, chemical and biological. The most effective management probably involves a combination of these. Mechanical techniques include hand-pulling, digging, root-cutting, use of weed eaters, axes, machetes, bulldozers, fire and flooding. Removal by hand is generally recommended for small infestations of saplings under 1-inch diameter. Root-cutting and bulldozing may be effective but are costly, labor intensive and may cause extensive damage to soils and lead to resprouting. Fire has been used with some success, but because saltcedars are fireadapted, they readily resprout after fire. Flooding can be used to control salt cedar if root crowns remain submerged for at least three months. For extensive infestations of saltcedar, chemical control has been shown to be the most effective method. Cautious use of herbicides aids in restoration of saltcedar infested sites by allowing repopulation by native plant species.

Systemic herbicides (e.g., those that kill the plant from the root up) are recommended for saltcedar management and application methods include foliar sprays, cut stump treatments, basal bark treatments, and aerial sprays. Because tamarisk usually grows in or adjacent to streams, wetlands and other waterways, it is important to use products registered for aquatic application. Fifteen insects are being investigated as potential biological control agents for saltcedar. Two of these, a mealybug (Trabutina mannipara) and a leaf beetle (Diorhabda elongata), have preliminary approval for release. Five others are being tested within the United States and an additional eight species are under study overseas. Final approval for release of the mealybug and the leaf beetle is pending resolution of concerns regarding their potential impact to the southwestern subspecies of the willow flycatcher (Empidonax trailii extimus), a federally endangered bird. In parts of its range where native willows, its natural nest trees, have been replaced by saltcedar, the willow flycatcher now utilizes it for this purpose. Concern is over the possibility that, due to the environmental damage caused by tamarisk, native plant species may not be able to replace it if the biological control agents succeed in eliminating it.

RECLAIMING THE PUERCO RIVER ARTICLE

A New Mexico conservation district works to rid the area of salt cedar.

When the McKinley Soil and Water Conservation District of Gallup, New Mexico, started exploring options for a comprehensive river restoration project, board members knew they had to start with the basics.

Although the board's ambitious plans included the restoration of a local river, the re-introduction of native plant and animal habitat and the eventual creation of a new wetland park area near Gallup, the board decided to first deal with a local infestation of salt cedar, an invasive, non-native plant that threatens water supplies throughout the West. "All over the West, there's a fevered debate over water," said Larry Winn, chairperson of the McKinley Soil and Water Conservation District in Gallup. "A large portion of that debate is focused on how water is being wasted. Communities like ours struggle with saving water, getting water and living in a desert, while still trying to maintain the natural beauty of plants, trees and rivers. Because of this debate, we realized that we needed to start with a salt cedar eradication project before we could bring back our river."

Salt cedar currently infests an estimated 1.5 million acres along waterways in the western United States. As a wily competitor for resources, a single acre of the invasive plant can suck up more than 2.8 million gallons of water in a single year, according to researchers from the Texas Cooperative Extension. Like hundreds of other communities throughout the western United States, Gallup's Rio Puerco river basin is heavily infested with the invasive tree.

RESTORING THE RIO PUERCO OF THE WEST

Gallup's Puerco River-known as the Rio Puerco of the West-is an ephemeral stream that runs through the city alongside the famed Route 66 highway. Like other ephemeral streams in arid northwest New Mexico and northeast Arizona, the Rio Puerco of the West usually only flows after occasional summer rains or when mountain snows melt in the spring. In general, it runs for just a few days or weeks each year, and lies dusty and barren the rest of the time.

Local water conservation boards, land managers and governmental bodies generally rely on one of several different methods to control salt cedar, including physical removal and through the application of herbicides that negatively affect salt cedar.

Today, the only place the river flows with any regularity is downstream of our water treatment plant, where treated effluent is released to mimic a perennial stream," Winn said. "We hope to change that by getting rid of salt cedar and restoring the river's natural flow using effluent from a new reverse-osmosis water treatment plant." At present, the Rio Puerco's natural streambed parallels an effluent ditch from Gallup's water treatment plant. The water conservation district plans to rehabilitate a two-mile section of the natural streambed, first by eradicating salt cedar along the stream, then by using treated effluent to mimic a perennial stream. The treated water will come from a state-of-the-art reverse osmosis water treatment plant scheduled to open in just over a year. When the entire project is complete, the board hopes to install a series of footbridges over the river and create a 10,000-acre nature park. "What we hope to do is bring the river back and reintroduce a certain interest in the proper scale of living in the desert," Winn said. "We want to create an appreciation for the natural world in our community. As we restore the river, we restore ourselves." Before such long-term objectives can be reached, however, the McKinley Soil and Water Conservation District is taking key steps to eradicate the negative effects caused by the region's salt cedar infestation.

ABOUT SALT CEDAR

Salt cedar, or tamarisk, is the common name for a species of exotic, invasive trees or shrubs inflicting significant damage on stream banks and waterways throughout the United States.

Originally imported from Eurasia to serve as a shade plant and erosion-prevention tool in arid regions of the western United States, salt cedar is today one of the most invasive nonnative plants in America. The species has invaded approximately 29,000 acres on 33 national wildlife refuges managed by the U.S. Fish and Wildlife Service.

An exotic plant that generally grows to around 20 feet tall, salt cedar has long taproots that allow it to intercept deep water tables and interfere with natural aquatic systems. Salt cedar can consume twice the water of native plants such as willow trees. Experts estimate that each salt cedar tree along the Pecos River in Texas uses 14,000 gallons annually-or 78 gallons per tree, per day. Like many invasive plant species, salt cedar has an exceptionally strong reproductive capability. A mature salt cedar plant can produce 600,000 seeds annually. After summer rains, seedlings can quickly colonize moist areas due to the constant availability of seeds. A mature salt cedar plant can also reproduce vegetatively, meaning its roots, cuttings and branches have the ability to resprout when detached.

Salt cedar trees can live 50 to 100 years and generally grow from four to 26 feet tall. A single tree can grow nine to 12 feet in a single season under good conditions. During a normal spring, it can grow as much as one foot per month. The weed is highly adapted to arid climates, thriving in salty, nutrient-poor soil. It is also extremely resistant to adverse environmental conditions such as drought or flooding. During a drought, the invasive weed survives by dropping its leaves and halting growth to conserve water and nutrients. Under flooding, it can survive immersion in water for up to 70 days. Salt cedar has invaded waterside ecosystems throughout the southwestern United States. Prevalent in Texas, New Mexico, Arizona and Utah, salt cedar also extends north to Colorado, Indiana, Kansas, Massachusetts, Missouri, Nebraska and Oklahoma. Well-established stands can also be found along waterways in Idaho, Montana, Oregon, South Dakota and Wyoming.

Highly Destructive Plant

Salt cedar poses a severe threat to the structure and stability of native plant communities throughout the United States. The invasive plant draws salt from deep in the soil and exudes it via glands in its leaves. As rain washes the salt off the leaves and into the soil, the resulting increase in salinity kills understory plants and inhibits surrounding native plants from growing.

From the southwestern United States up through Montana, stands of salt cedar have severely limited the number of germination sites suitable for native waterfront species such as willow and cottonwood. The foliage and flowers of salt cedar provide little food value for native wildlife species, which depend on nutrient-rich native plant species.

Numerous studies have shown that stands of salt cedar can seriously degrade natural habitats. Salt cedar infestations can displace or replace native plant communities, degrade wildlife habitat and contribute significantly to the decline of many native species, including several threatened or endangered species. Along the lower Colorado River, for example, an influx of salt cedar into native habitats seriously threatens nine bird species. According to a report by the California Department of Fish and Game, two native bird species have been completely eliminated from the area. Seven other species are in serious decline, including the elf owl, Bell's vireo, Gila woodpecker, gilded northern flicker, vermilion flycatcher, summer tanager and western yellow-billed cuckoo.

Because salt cedar branches can form dense thickets many feet wide, the plant can impede wildlife access to water sources. In the waterways themselves, salt cedar adversely affects many fish species by creating narrow, homogenous stream habitats and reducing the number and variety of insects used as a food source. Cattle and other livestock usually don't prefer salt cedar as a food source because it quickly grows taller than they can reach. Salt cedar can actually reduce a rangeland's livestock capacity by displacing forage grasses, increasing soil salinity, raising the incidence of fires and using groundwater or irrigation water that could otherwise be available to grow forage plants.

Bringing a River Back from the Desert

In Gallup, Winn said the idea to remove salt cedar came about after he saw how much affected ecosystems can improve after the plant is brought under control. Studies show that removing salt cedar infestations can improve both quantity and quality of water, help to rejuvenate underground springs, increase stream flow and return wetlands to their intended state.

"Last year, I attended a workshop for stream health in Santa Fe, New Mexico, near the Santa Fe River-which is in the same situation as our Rio Puerco," Winn said. "It generally doesn't flow except for annual rains and the introduction of treated effluent. The city of Santa Fe got permission to restore a streamside riparian forest area that once existed there, and it just looked wonderful. After seeing their remarkable success, I came back to Gallup and started talking it up around town. People here were very receptive and interested."

Because salt cedar is so well adapted to harsh conditions, controlling the plant can often be difficult and costly, depending on the measures used. Local water conservation boards, land managers and governmental bodies generally rely on one of several different methods to control salt cedar, including physical removal-either by hand or through mechanical means-and through the application of special targeted herbicides that negatively affect salt cedar.

The McKinley Soil and Water Conservation District in Gallup opted to use a newer tactic: the application of a new, smart herbicide developed by BASF Corporation. Unlike most other herbicides on the market, Arsenal herbicide from BASF can provide effective salt cedar control without negatively affecting most grasses, forbs and legumes-which serve as prime wildlife forage and habitat. The herbicide, which is typically applied in late summer to fall for maximum effectiveness, works by inhibiting three key amino acids essential to plant growth. The herbicide translocates to the roots and shoots and inhibits a specific enzyme that causes the salt cedar to stop growing and slowly die as its nutrient reserves are exhausted.

In Gallup, a recent salt cedar-eradication demonstration project used helicopters, GPS guidance and other tools to target herbicide delivery specifically to a two-mile stretch of salt cedar-infested riverbank areas. The treated areas are owned by the city of Gallup and three other landowners-all of which agreed to participate in the salt cedar eradication test project. The parties involved agreed to the project in part after learning that Arsenal herbicide breaks down quickly in the soil, allowing desirable grasses, forbs and legume species to germinate and repopulate a treated site. Because it is considered a low-volume herbicide, it provides more control with less chemical load on the environment, compared to other herbicides.

Biological Control Efforts

Scientists are currently investigating the use of non-native insects as potential biological control agents for salt cedar in the United States. Two insects, Trabutina mannipara (a mealybug from Israel) and Diorhabda elongate (commonly known as the Asian leaf beetle), are natural predators for salt cedar and have been approved for release by the U.S. Department of Agriculture. In 2001, researchers released Asian leaf beetles in field sites throughout the West. However, it will likely be several years before the leaf beetle's success can be seen. According to scientists, even if the beetle is effective, the insect only provides around 75-80 percent control of salt cedar and may take as much as 10 years before results are noticeable.

Salt cedar eradication gains momentum in other communities. And, since the insects rely on salt cedar as a food source, they are unlikely to control all salt cedar plants. In a best-case scenario, the Asian leaf beetle would serve as means of damaging-but not eradicating-stands of salt "I think bringing back the river is a metaphor for bringing life back to people and our region. Restoring it can help improve the quality of life in our community. " -Larry Winn, chairperson of the McKinley Soil and Water Conservation District in Gallup, New Mexico

Several communities in the West and Southwest have used Arsenal with significant success. In the 1990s, the U.S. Fish and Wildlife Service began using BASF's Arsenal as part of an extensive, longrunning campaign to eradicate salt cedar in the Bosque del Apache National Wildlife Refuge near Socorro, New Mexico. Aimed at rehabilitating salt cedar-infested floodplains and restoring diverse riparian habitats, the program used Arsenal in a number of different combinations to test for the most effective application methods. A study conducted by researchers from New Mexico State University found that a helicopter application of Arsenal alone provided 90 percent or greater control of salt cedar. Similarly, as part of a salt cedar-control program still underway in Texas, Arsenal is being applied with significant success in key salt cedar-infested areas along the Pecos River. Between 1999 and 2002, the Texas Cooperative Extension used Arsenal to treat approximately 6,341 acres of salt cedar along 120 miles of the Pecos River.

After just three years of treatments, salt cedar control efforts have salvaged just under 12 billion gallons of water-roughly enough water to serve a city the size of Lubbock, Texas, for a year. No significant re-invasion of salt cedar has occurred in the Arsenal herbicide-treated areas and annual water savings in 2003 are estimated to increase to 15.9 billion gallons.

HOPEFUL FOR THE FUTURE

Although salt cedar continues to aggressively attack the environment by infesting millions of acres of land and displacing native habitat, the development of innovative solutions, such as Arsenal, are finally enabling landowners and managers to take firm control of the problem. The result is the ability to reclaim precious land and water resources-saving ecosystems, wildlife and livelihoods. In Gallup, the McKinley Soil and Water Conservation District hopes that its first demonstration test using Arsenal will encourage other communities downstream in New Mexico and Arizona to consider similar eradication tactics and river-restoration programs.

"There are many communities downstream that have water treatment plans along the Rio Puerco and the Little Colorado," Winn said. "If we can get other states and water boards interested in salt cedar eradication and high-tech water treatment systems, we can create significant stretches of river that run year-round."

Winn said he hopes the demonstration project will renew interest in salt cedar control and increase the likelihood of larger application projects.

"If we can bring back this river one stretch at a time, we have, in effect, brought back life within ourselves," Winn said. "I think bringing back the river is a metaphor for bringing life back to people and our region. Restoring it can help improve the quality of life in our community."

SIBERIAN ELM ULMUS PUMILA L.

Native Range: Northern China, eastern Siberia, Manchuria and Korea

Description: Siberian elm is a fastgrowing tree in the elm family (Ulmaceae) distinguished by small toothed leaves about 1-2½ in (3-7 cm) long and half as wide, and pointed at the tip. Unlike other elms, the leaf base is usually symmetrical, forming a nearly even "V". Leaves are smooth and dark green above, paler and nearly hairless beneath, and alternate from side to side along twigs.

Mature trees reach a height of 50-70 ft. (16-22 m.), with a round crown of slender, spreading branches. The bark is rough, gray or brown, and shallowly furrowed at maturity. Twigs are nearly hairless with small, blunt buds. Flowering occurs in the springtime. The small greenish flowers lack petals and occur in drooping clusters of 2 to 5. After flowering, a single seed forms in the center of each smooth, flattened, circular, ½ in (10-15 mm) wide fruit.

Other species of elms (Ulmus) and the close relative Zelkova, especially younger plants, look similar to Siberian elm. Some may even confuse it with choke-cherry (Prunus serotina) and hackberry (Celtis sp.). The native slippery elm and American elm typically have leaves that are greater than 3 in (7.3 cm) long, with unequal heart-shaped leaf bases, and leaf margins with double teeth.

Ecological Threat: Dry to mesic prairies and stream banks are vulnerable to Siberian elm invasion. Thickets of seedlings soon form around seed-producing trees, bare ground areas, animal and insect mounds, and other disturbed areas. Wind carries seed to distant areas where new colonies can form. This tough exotic survives under conditions not easily tolerated by other species, allowing it to take advantage of open ground and resources otherwise used by native plants.

Fast growing seedlings of Siberian elm quickly overtake native vegetation, especially shade-intolerant species. This often leads to invasion by additional weedy species, compounding the problem.

Distribution in the United States: Siberian elm is known to occur in 43 states and reported to be invasive in natural areas in 25 states (AZ, IA, ID, IL, IN, KS, KY, MA, MD, MI, MN, MO, NE, NM, NV, OH, OK, OR, PA, TX, UT, VA, WA, WI, WV).

Habitat in the United States: Dry and mesic prairies and areas along stream banks in Minnesota and forested areas and high elevations in Arizona.

Background: Siberian elm was introduced to the U.S. in the 1860's for its hardiness and fast growth in a variety of moisture regimes and habitats, including droughts and cold winters. It is resistant to Dutch elm disease. This elm continues to be promoted, especially in the Great Plains, in spite of weak limbs and susceptibility to insect attack.

Biology and Spread: Seeds are produced early in spring and spread by the wind. Germination rate is high and seedlings soon establish in the bare ground found early in the growing season.

Current Management Approaches: For long term management of Siberian elm, reduction of seed sources is essential

Chemical: To avoid resprouts after cutting or girdling, cut stumps may be treated with systemic herbicides such as glyphosate (e.g., Roundup) and triclopyr (e.g., Garlon). After spring sap flow ceases and during the growing season, Minnesota DNR State Parks Southern Region Resource Management applies 4 parts water to 1 part glyphosate (based on 41% active ingredient glyphosate concentrate) with a hand sprayer to cut stumps.

The entire stump should be saturated with the herbicide to achieve the most effective control. Garlon 4 (triclopyr ester formulation) can be applied as a cut stump or basal bark treatment. For basal bark, apply a 20-percent solution in horticultural oil (2.5 quarts per 3-gallon mix) with a penetrant (check with herbicide distributor) to young bark as a basal spray in winter (January to February) or summer-fall (June to October). Cut stems can be immediately treated with glyphosate herbicide as a 20-percent solution (2.5 quarts per 3-gallon mix) in water with a surfactant.

Manual: During the growing season, seedlings can be hand pulled and small trees carefully removed by a grub hoe or weed wrench. Trees girdled in mid-May to early July will die over 1-2 years without sprouting if cut properly.

To remove a band of bark from the wood, make two parallel cuts 3-4 inches apart, then knock bark off with a blunt object such as the back of an axe head or dull end of a girdling bar. The xylem must remain intact; if girdled too deeply the tree will respond as if cut down and will resprout. On sites with few seed sources, the large trees can be cut down and resprouts trimmed as needed.

Other: A regular regime of prescribed burning in fire-adapted communities will kill seedlings.

CASTOR BEAN RICINUS COMMUNIS

Family: Euphorbiaceae (spurge family)

Common Names: Castor bean, castor oil plant, palma Christi, wonder tree Description.

In frost-free areas, castor bean is an evergreen herbaceous or semi woody large shrub or small tree that gets up to 40 ft

(12 m) tall and 15ft (4.6 m) wide. In the tropics, it can have a trunk that is woody near the base and up to a foot in diameter. Elsewhere, castor bean plant grows as an annual that can get 8-15 ft (2.4-4.6 m) tall in a single growing season. This is a fast growing, suckering, and colony forming plant with decidedly tropical looking foliage.

They tend to grow straight up at first, developing branches only later in the season (and in subsequent years for plants that live that long). The huge leaves are palmate, with 5-11 deeply incised lobes. They are glossy purplish or reddish-green and 12-30 in (30-76 cm) across, with long petioles (leaf stems). The stems are watery juicy and reddish or purplish too.

The inflorescence is not particularly showy; small, 0.5 in (1 cm) wide greenish yellow flowers are borne in fat spikes 8-18" tall near the tops of the stems. Female flowers are on the top half of the spike and have conspicuous red stigmas (the parts that receive the pollen).

The male flowers on the lower half of the spike have conspicuous yellow anthers (the parts that give off the pollen). The female flowers are followed by reddish brown egg-shaped capsules, about an inch long, thickly covered with soft flexible spines. Each capsule contains three seeds that look like fat swollen dog ticks and are deadly poisonous.

There are several named cultivars, including some grown commercially for oil production, and this sampling of ornamentals: 'Carmencita' has bronzy red leaves and bright red female flowers. 'Impala' is small, 4-5 ft (1.2-1.5 m) tall and has red leaves that age to reddish purple. 'Sanguineus' has blood-red stems and leaves. 'Gibsonii Mirabilis' is dwarf, only 4' tall, and has dark red leaves and stems. 'Zanzibarensis' is taller and has larger leaves that are green with white veins.

Location: Castor bean was originally native to northeastern Africa and the Middle East. It has escaped cultivation and become naturalized as a weed almost everywhere in the world that has a tropical or subtropical climate. Castor bean grows wild on rocky hillsides, and in waste places, fallow fields, along road shoulders and at the edges of cultivated lands.

The Problem: Castor oil plant spreads over sandy soil areas, creekbanks, and gullies. This can lead to a significant loss of prime grazing land. The seeds of castor oil contain ricin, a poison which is extremely toxic to livestock and humans. Leaves have a lesser amount of toxin. Symptoms of poisoning in animals usually do not appear for a few hours or several days.

Seeds cause gastro-intestinal disorders; leaves tend to cause neuro-muscular disorders. Poisoning in livestock is rarely reported though, as castor oil plant is seldom grazed by stock when other pasture plants are available. Also, small amounts of the plant will induce immunity to poisoning.

Control: Individual plants or small infestations may be removed by cultivation or hand-pulling. Broad-scale infestations may require spraying with herbicides to control the plant.

Bibliography

Angel, Bradley: *Toxic Threat to Indian Lands: A Greenpeace Report*, San Francisco, Greenpeace, 1991.

Bailey, Conner, Kelly Alley, Charles E. Faupel, and Cathy Solheim: *Environmental Justice: Issues, Policies, and Solutions*, Washington, D.C., Island Press, 1995.

Berry, Brian L.: *The Social Burdens of Environmental Pollution: A Comparative Metropolitan Data Source*, Cambridge, Ballinger Publishing Co., 1977.

Bryant, Bunyan and Paul Mohai: *Race and the Incidence of Environmental Hazards: A Time for Discourse*, Boulder, Westview Press, 1992.

Bullard, Robert D.: *Unequal Protection: Environmental Justice and Communities of Color*, San Francisco, Sierra Club Press, 1994.

Burch, William R. Jr.: *Human Ecology: An Environmental Approach*, Belmont, Duxbury Press, 1994.

Campbell, R. R., and Wade, J. L.: *Society and Environment: The Coming Collision*, Boston, Allyn and Bacon, 1972.

Dasmann, R.F.: *Environmental Conservation*, New York, Wiley, 1972.

Daubenmire, R. F.: *Plants and Environment*, New York, Wiley, 1947.

Erickson, Kai: *A New Species of Trouble: Explorations in Disaster, Trauma, and Community*, New York, W.W. Norton, 1994.

Freeman, A. Myrick: *Environmental Quality Analysis: Theory and Method in the Social Sciences*, Baltimore, John Hopkins University Press, 1972.

Gay, Kathlyn: *Saving the Environment: Debating the Costs*, New York, Franklin Watts, 1996.

Gelobter, Michael: *The Distribution of Air Pollution by Income and Race*, Berkeley, University of California, 1989.

Gerrard, Michael B.: *Whose Backyard, Whose Risk: Fear and Fairness in Toxic and Nuclear Waste Siting*, Boston, MIT Press, 1994.

Hill, David: *The Quality of Life in America; Pollution, Poverty, Power, and Fear*, New York, Holt, Rinehart and Winston, 1973.

Hofrichter, Richard: *The Theory and Practice of Environmental Justice*, Philadelphia, New Society Publishers, 1993.

Johnson, Rebecca L.: *The Greenhouse Effect: Life on a Warmer Planet*, Minn., Lerner Publications, 1993.

Kruvant, W.J.: *People, Energy, and Pollution*, Cambridge, Ballinger Publishing, 1975.

Mandelker, D.R.: *Environment and Equity: A Regulatory Challenge*, New York, McGraw Hill, 1981.

Middleton, Nick: *Atlas of Environmental Issues*, New York, Facts on File, 1989.

Sachs, Aaron: *Eco-Justice: Linking Human Rights and the Environment*, Washington, World Watch Institute, 1995.

Sontheimer, Sally: *Women and the Environment: Crisis and Development in the Third World*, New York, Monthly Review Press, 1991.

Vallette, J.: *The International Trade in Wastes: A Greenpeace Inventory*, Washington, Greenpeace, 1989.

Wenz, Peter S.: *Environmental Justice*, Albany, State University of New York Press, 1988.

Zupan, J.M.: *The Distribution of Air Quality in the New York Region*, Baltimore, Johns Hopkins University Press, 1973.

Index

R

S

T

V

W

❑❑❑